Better Takeoffs
& Landings

TAB
PRACTICAL
FLYING SERIES

Other Books in the TAB PRACTICAL FLYING SERIES

Aviator's Guide to GPS—2nd Edition *by Bill Clarke*

Cockpit Resource Management: The Private Pilot's Guide
by Thomas P. Turner

Handling In-Flight Emergencies *by Jerry A. Eichenberger*

**Pilot's Guide to Weather Reports, Forecasts,
and Flight Planning—2nd Edition** *by Terry T. Lankford*

The Art of Instrument Flying—3rd Edition *by J. R. Williams*

Avoiding Mid-Air Collisions *by Shari Stamford Krause, Ph.D.*

Cross-Country Flying *by Jerry A. Eichenberger*

Flying in Adverse Conditions *by R. Randall Padfield*

Multiengine Flying *by Paul A. Craig*

Weather Patterns and Phenomena: A Pilot's Guide *by Thomas P. Turner*

Better Takeoffs & Landings

Michael C. Love

TAB Books
Division of McGraw-Hill, Inc.
New York San Francisco Washington, D.C. Auckland Bogotá
Caracas Lisbon London Madrid Mexico City Milan
Montreal New Delhi San Juan Singapore
Sydney Tokyo Toronto

pbk   1 2 3 4 5 6 7 8 9 FRG/FRG 9 9 8 7 6 5
hc   1 2 3 4 5 6 7 8 9 FRG/FRG 9 9 8 7 6 5

Library of Congress Cataloging-in-Publication Data

Love, Michael Charles.
    Better takeoffs & landings / by Michael Charles Love.
        p.   cm.
    Includes bibliographical references and index.
    ISBN 0-07-038805-9 (h).   ISBN 0-07-038806-7 (p)
    1. Airplanes—Take-off. 2. Airplanes—Landing. I. Title.
TL711.T3L68   1995   95-13583
629.132'5212—dc20   CIP

Acquisitions editor: Shelley Chevalier
Editorial team: Robert E. Ostrander, Supervising Editor
                Sally Anne Glover, Book Editor
Production team: Katherine G. Brown, Director
                Toya L. Warner, Computer Graphic Artist
                Wanda S. Ditch, Desktop Operator
                Lorie L. White, Proofreading
                Jodi L. Tyler, Indexer
Design team: Jaclyn J. Boone, Designer   TPFS
                Katherine Stefanski, Associate Designer   0388067

To Dottie, Ryan, Emily, and Mr. Young.

Acknowledgments

Thanks to William LeGore, Airport Operations Supervisor at Dane County Regional Airport, for his aid with pictures taken for the book. Thanks also to the control tower staff at Dane County Regional for their cooperation during the session. Pictures were taken by Matthew Love, who braved Wisconsin's winter cold to get them. Finally, a big thanks to Lee Johnson for proofreading the manuscript and for the use of the Pitts S2-B. Thanks to the staff of Wisconsin Aviation, Four Lakes, also at Dane County Regional, for all their cooperation.

Contents

Introduction   xi

1 Flight safety topics   1
Pilot proficiency   1
Aircraft preflight   3
Weather   7
Summary   11

2 V-speeds   13
V-speeds   13
The airspeed indicator   18
V-speed in the pattern   20
Summary   20

3 The traffic pattern   23
Initial radio call approaches   23
Departures   25
Left and right patterns   26
Ground markings   28
Runway markings   30
Pattern entry   34
Pattern exit   40
Aircraft separation   42
The busy pattern   46
Pattern wind correction   47
Summary   48

4 Landing techniques   51
Landing checklist   51
Airspeed/power control   54
Picking the touchdown point   55
Vision during the flare   60
Height judgment during flare   61
Flaring/touchdown   62
Ground effect   68
Unusual landing situations   71
Flat approaches   74
Groundloops   75

Night landings   76
Wet runways   81
Summary   81

5 Normal takeoffs and landings   83
Takeoffs   84
Landing   95
Summary   100

6 Short-field takeoffs and landings   103
Performance charts   104
Takeoff charts   105
Landing charts   109
Takeoff   113
Landing   119
Summary   127

7 Soft-field takeoffs and landings   129
Ground effect/stall   130
Takeoff   131
Landings   139
Summary   144

8 Crosswind takeoffs and landings   147
Crosswind taxiing   148
Crosswind control use   150
Crosswind airspeeds   151
Takeoff   153
Landing   159
Summary   166

9 Winter flying   167
Preflight   168
Engine preheat   168
Taxiing   176
Takeoff run   178
Landing rollout   178
Summary   180

10 **Abort procedures**   **183**
Takeoff aborts   184
Landing abort (go-around)   187
Pilot indecision   192
Summary   192

11 **Emergency-landing procedures**   **195**
Pilot scan   196
Warning signs   197
Engine loss on takeoff   200
Engine loss on approach to landing   204
Rugged-terrain landings   207
Retractable-gear problems   213
Flat-tire landing   214
Downwind landings   215
Summary   217

12 **Conclusion**   **219**
Where to go from here   219
Flight safety   223
Summary   225

Bibliography   **227**

Index   **229**

About the author   **235**

Introduction

FLYING HAS FASCINATED ME SINCE CHILDHOOD. I WOULD RIDE MY BIKE TO the local airport to watch airplanes take off and land, dreaming of the day when I would become a pilot. I eventually realized my dreams, soloing when I was 16 and going on to earn my commercial, instrument, and flight instructor ratings. To this day I am as captivated by flight as I was as a young boy. This book is an outgrowth of that interest in flying and a desire to see pilots be as safe and competent as possible. No matter who we are, or what our level of flight experience is, we can always improve our flying skills.

This book discusses techniques associated with different types of takeoffs and landings. Pilots who hold a minimum of a private pilot certificate will find this a helpful, instructive manual. However, it could also be used by student pilots in conjunction with their standard flight-training syllabus.

All pilots must work at maintaining their flying abilities, and this book is intended to aid them in performing proficient takeoffs and landings. These two areas are where many general-aviation accidents take place due to the close proximity to the ground and slower flying speeds. The overall impression our passengers have of their flight can also be directly related to how smoothly we perform these tasks. When my passengers comment about a smooth flare and a gentle touchdown, I know I have done a good job.

As pilots we should be able to take off and land smoothly and accurately on a consistent basis. We must also be able to adapt to varying weather conditions when we fly. In the following chapters you will find reminders of many of the takeoff and landing topics that were covered during your flight training, and possibly some areas that were not. Techniques for normal, short-field, soft-field, and crosswind takeoffs and landings will be given an in-depth review. Use of the traffic pattern and emergency procedures are also covered. Additionally, there is a discussion of weather factors and how to deal with them to achieve the level of performance that makes us safer pilots.

Throughout the book, examples are used to illustrate the points being made. When you are through, you will have a solid understanding of the accepted methods for executing takeoffs and landings in varying weather conditions in a safe, consistent manner.

Many pilot-training manuals only concentrate on tricycle-gear aircraft. Having had the opportunity to fly and instruct in tailwheel aircraft, I feel this is an area that also needs to be addressed. Where applicable, this book will include takeoff and landing information for tailwheel aircraft. I have had some of my own students try to remove runway lights during landings, and that has reinforced the need to include specifics about standard-gear aircraft.

When you go out to practice takeoff and landing maneuvers as presented, remember that you should have a qualified flight instructor accompany you until you are proficient. Safety is always the highest priority. Becoming an FAA statistic and trying to improve your flying ability are contradictory tasks, so take advantage of the safety and additional tips an instructor can bring as you practice. Also, review your aircraft's operations manual for specifics on items such as airspeeds and flap and power settings for each maneuver.

The first chapter discusses safety topics that include pilot proficiency, aircraft preflight, and weather factors. Pilots must fly at regular intervals and practice specific maneuvers to be proficient in them. Takeoffs and landings are no exception. This chapter discusses what pilots must do to reach and maintain an acceptable skill level. The second topic is a review of aircraft preflight activities. Who among us has not skipped steps in a preflight to get in the air sooner? An example of a thorough, fast preflight is given step-by-step coverage. Finally, the chapter concludes with weather factors. Included among them are wind shear and its effect, and how temperature and humidity influence aircraft performance.

Chapter 2 covers V-speeds, which are central to our ability to fly an airplane safely. Knowledge of these speeds as related to takeoffs and landings is essential. Using the aircraft operations manual to find them is a necessary task we must perform for every aircraft we fly. We also need to know how to use the V-speeds. For instance, do you know what multiples of V_{SO} the FAA recommends for use on downwind and final? Having a working knowledge of V-speeds is fundamental to being a safe pilot.

Traffic pattern use is covered in chapter 3. Smooth entry into and exit from an airport's traffic pattern not only increases the safety of our flight, but also the safety for other aircraft in our vicinity. Review of aircraft separation through continual visual scanning and use of the radio are included in the chapter.

Chapter 4 discusses landing techniques that include how to pick your landing spot while on downwind, vision during the flare, and height judgment. The importance of airspeed control and maintaining a constant glidepath are also reviewed. Tailwheel topics include wheel and three-point landing techniques.

Chapters 5 through 8 are dedicated to the specific techniques of normal, short-field, soft-field, and crosswind takeoffs and landings. Most general-aviation pilots today use long, wide, paved runways and rarely land on less-improved runways. Soft ground, tall grass, or snow can make a seemingly adequate strip suddenly too short if the proper techniques are not used. Within each of these chapters, the correct use of controls and airspeeds will be given detailed coverage. The impact of various weather conditions will also be included. Ground effect plays an important part in soft-field takeoffs, and unless you practice them more regularly than your biennial flight review, you might find them difficult to do proficiently when it really counts.

Winter flying is the topic of chapter 9. This might be of special interest to those southern pilots who are infrequently exposed to the frigid weather that northern pilots have become accustomed to. The thin layer of frost on your wings in the morning can be as dangerous as an accumulation of snow. How and why frost and snow on an aircraft can affect takeoff and landings will be reviewed, in addition to cold weather engine operation and taxiing on snow or ice.

Chapter 10 is concerned with abort procedures for both takeoffs and landings. Execution of go-arounds is discussed in detail. In particular, the results of hesitating before committing to a go-around and how to avoid this potential problem are reviewed.

Emergency landing procedures are discussed in chapter 11. The average pilot is exposed to these every two years during the biennial flight review when the check pilot pulls the throttle back and states that the engine has quit. It is important to think about this on a continual basis. This chapter also covers emergency landing procedures for open areas and rugged terrain. Many pilots who have lost an engine shortly after takeoff have unsuccessfully tried to perform a 180° turn to land on the runway they took off from, all too frequently with fatal results. The FAA's recommended minimum safe altitude for a return to the runway and the rationale behind it will be analyzed.

The book concludes with a general review of skills maintenance and flight safety. There is no substitute for "doing." This means getting out there and flying and applying the information and techniques you have been taught on a regular basis. Flying is not just moving your hands and feet to make the aircraft take off and land. Flying is applying sound, rational judgment to increase the safety of your flights.

To maintain "judgment proficiency," you need to continually review publications that the Federal Aviation Administration makes available, attend safety seminars, review the Federal Aviation Regulations and, yes, read books covering aviation-related subject matter. By continuing your aviation education, you will become a safer, more competent pilot.

I hope that as you read the following chapters you take them as they are intended. Being a pilot should be an enjoyable activity, but the responsibility we have as pilots to ourselves and to our passengers is quite serious. Even if you know every fact presented in this book, you should come away from it with an appreciation of the pleasure and responsibility that being a pilot gives each of us.

1
Flight safety topics

FLIGHT SAFETY IS ESSENTIALLY WHAT THIS BOOK IS ALL ABOUT. AS YOU practice your takeoffs and landings and apply what you have learned, you will not only improve your flying abilities, but you will also increase your level of flight safety. Before we get to specific details concerning takeoffs and landings, we need to cover three topics: pilot proficiency, aircraft preflight, and weather factors. Each of these subjects pertains to flight safety and needs to be reviewed.

When you have completed this chapter, you will have a better understanding of what pilot proficiency is and how to maintain your proficiency. How often you fly and the type of flying you do does make a difference. This chapter also covers aircraft preflights and what you need to be aware of as you do them. You will learn what wind shear is and how and why temperature and humidity affect an aircraft's performance. Let's begin with pilot proficiency.

PILOT PROFICIENCY

Part 61.57 of the Federal Aviation Regulations (FARs) states that no person may act as the pilot in command of an aircraft carrying passengers unless he or she has made 3 takeoffs and landings in the same category and class of aircraft within the preceding 90 days. While the Federal Aviation Administration (FAA) might consider this a minimum proficiency requirement for carrying passengers, no pilot can perform at peak levels by meeting only the minimums of this regulation. Taken at face value, performing 3 takeoffs and landings every 90 days means that in one year's time, you need to fly only 5 times, doing 3 touch and goes, for a total of 15 takeoffs and landings!

To do anything well, you need to practice doing it on a regular basis, and meeting the FAA minimum for pilot proficiency is not sufficient. There have been periods of time in the past when, because of work and school schedules, I was not able to fly as often as I wanted. When I did fly during those times, I could tell that my level of performance had fallen below what I considered to be acceptable standards. I was meeting the minimum requirements, but I wasn't pleased with my flying. I can see and feel improvements when I fly on a regular basis. For example, flying for skydivers, I might make 8 to 12 takeoffs and landings, in varying weather conditions, in a single day. This high frequency of repetition over a short period of time shows a definite improvement in my takeoffs and landings.

What is proficient?

For the purposes of this book, I want to define takeoff and landing proficiency as the ability to consistently perform takeoff and landing maneuvers in complete control of the aircraft in accordance with the procedures outlined in the aircraft's operations manual. This definition of proficient is a minimum that each of us needs to strive for, but it is not the end goal. Note that proficiency as defined means that you might be proficient in one make and model aircraft, but not in another. A Cessna 172 has a different set of landing characteristics than a Citabria. While you might be able to "grease it on" in the 172 you fly every weekend, you are likely to end up bouncing all over the runway if you try to fly a Citabria the same way. Maintaining proficiency allows us to fly the aircraft safely, but it still leaves room for improvement.

To improve how well we fly, we need to be more than just proficient. We need to fly more frequently and make a conscious effort to critique ourselves during and after each takeoff and landing. For instance, on takeoff roll are you using correct crosswind control compensation? During climb, are you keeping the ball in the turn and bank indicator centered by using right rudder, or are you letting P-factor yaw the aircraft? In the pattern, are you compensating for wind, or are you drifting in relation to the runway? During your landing descent, are you using left rudder to keep the ball centered? Are your airspeeds steady, or do they fluctuate? Do you frequently make throttle adjustments to control your glidepath? These are just a few of the factors each of us needs to monitor if we want to become more than just an average pilot.

Flight frequency

How often a pilot needs to fly to maintain proficiency will vary for each individual. Some pilots might be able to fly an hour a week and maintain a satisfactory level of performance. Others might require more or less practice. In my experience, if you fly less than once every two weeks, you are probably not able to remain proficient according to the previous definition. That does not mean you are unable to fly, but it is likely you could do a smoother, more consistent job.

For most of us, there is a question of how often we can afford to fly. If you own your own aircraft, then you can probably fly on a weekly basis. Those who rent might have a tougher time affording to fly more often than weekly. As with all things that affect our budgets, you'll need to decide how often you can afford to get into the air to practice. The key is to use your time wisely to get the most practice you can. With that said, let's assume you can fly on a weekly basis. To improve your level of performance, you need to practice several takeoffs and landings each week. How many? Again, this will depend on the individual, but initially I would recommend about six touch and goes a week. By performing this many each week, you will begin to notice the little things that affect how you fly. A number of these factors were previously noted, and these and others will be covered in greater detail throughout the book. When you become aware of each of these factors and begin to compensate for them, you will be pleased with how consistent, smooth, and accurate your takeoffs and landings become.

As your level of performance improves, and compensating for all the factors becomes more consistent, you will be able to reduce the amount of time you need to practice. But don't fall into the trap of complacency and let your newly acquired skills deteriorate. Always monitor your performance and practice as often as necessary to stay at the highest level you can.

Proficiency and flight safety

It should be fairly obvious that proficiency and flight safety are closely tied. A proficient pilot knows how to safely fly the aircraft and avoids situations that can endanger the pilot, the passengers, or other aircraft. Pilots who are not proficient can find themselves in over their heads, and the result is all too often fatal. Each year pilots and their passengers are killed because of "pilot error." By maintaining proper skill levels and working to improve levels of performance, each of us can increase the safety and enjoyment of each flight.

AIRCRAFT PREFLIGHT

The next area we will cover is the subject of the aircraft preflight. Prior to each flight, we should perform a thorough walk-around of the aircraft. It is not possible to completely discuss the topic within the scope of this chapter, but I want to briefly cover a few areas concerning preflights that each of us should be familiar with.

What pilots need to know

No one expects a pilot to pull inspection panels, cowling, and fairings prior to each flight and perform an inspection that rivals the annual given by your favorite aircraft mechanic.

But during your preflight you need to look for certain things that can be signs of potential problems. Oil stains in the cowling, bent engine mounts, popped or deformed rivets, and buckled skin each have a story to tell the observant pilot. However, if you don't look for these signs, you could be flying an aircraft that has serious mechanical or structural problems. The following sections step through a preflight inspection example. Remember that the aircraft you fly will likely be somewhat different in its configuration, and you should consult its operations manual for preflight specifics.

Preflight inspection: Example

I fly a variety of aircraft, and for consistency I like to always begin the preflight at the same point, working my way around the airplane until I am back at the starting point. For the purpose of this example, I am going to use the Piper Tomahawk trainer. The key to this example is what I am looking for. Your aircraft might have a different order for the preflight inspection steps, but you should be looking for the types of things I am outlining here.

Preflight: The start. I begin the preflight inspection by opening the cockpit and verifying that the mags and master are off, then I lower the flaps. I then exit the cockpit and move to the right side of the cowl. This is where I start the inspection, and I will also finish at this point. I look over the external portion of the cowl, looking for missing screws and cracks. While looking at the underside of the cowl, I look for excessive or unusual oil or hydraulic fluid stains. I then open the right side of the cowl and check the oil level, verifying that it is not below minimum levels. I check the plug wires and plugs, giving each a slight counterclockwise twist to make sure they are not loose. Next I look over the rest of the engine, fire wall, and engine mounts. Again, I am looking for oil or fluid stains that might indicate a leak.

I look for bent engine-mount tubing, cracked welds, and missing bolts, nuts, and cotter keys. I also look for worn engine-mount biscuits. I inspect the exhaust manifold for missing nuts or exhaust stains that might show signs of a deteriorated gasket or cracked manifold. As I glance around, I verify there are no wires or cables that appear to be hanging free, and that there are no foreign objects in the engine compartment. I then secure the cowling and move to the front of the aircraft.

Preflight: The front of the aircraft. At the nose I inspect for cracks in the cowling and spinner. I also verify that the screws are in the spinner and give it a slight sideways push. If the spinner backing plate is cracked or loose, it might show up as play in the spinner as I wiggle it. I then inspect the propeller for nicks and cracks. Any nicks should be dressed by an aircraft mechanic to prevent them from becoming stress fracture points, which might become cracks in the propeller. If you do find a crack in a propeller, don't fly. There have been cases where cracks caused a portion of the propeller to separate, the propeller being so out of balance that it shook the engine off the engine mount in a matter of seconds. It goes without saying that not having the engine seriously affects your center of gravity and the flight characteristics of the airplane.

I also give the propeller a tug to make sure there is no play in it. A word of caution at this point: if you ever need to turn the propeller, always turn it opposite its direction of normal rotation. Magnetos generate the spark that ignites the fuel. They are grounded when the ignition key is in the off position, preventing them from generating an electrical current. However, if there is faulty wiring, the magneto could produce a spark when the propeller is turned in the direction of normal rotation. This could result in the engine inadvertently firing, turning the propeller you are holding. Turning opposite the rotation prevents the magneto from generating a spark.

While in front of the aircraft, I look over the nose wheel, making sure the tire is not worn beyond acceptable levels and that all bolts in the wheel, steering linkage, and torque link are present and secured. I then verify the nose wheel strut is inflated. After completing the nosewheel inspection, I look at the ground under the nose of the aircraft and hunt for any stains or puddles of fluid. At this time I move to the left side of the cowling.

Preflight: Cowling left side. On the left side of the Tomahawk, I perform a repeat of the right-side inspection. The only difference is that, instead of checking the oil, I check the brake fluid level. Other than that, I again check for missing screws and bolts, fluid leaks, bent or cracked engine-mount tubing, and the other items that were previously mentioned. After completing the inspection of the left side, I secure the cowl and drain fuel from the fuel drain.

I check the fuel for water at the bottom of the fuel drainer. I look for particles suspended in it and for the correct color. Whenever possible I try to look at the fuel against a white background. If you hold the strainer up to the sky to verify that it is the right color—for example, blue for 100LL (low lead)—the color of the sky might make clear fluid look blue. Remember that if you mix two grades of avgas, they become clear. Today many airplanes have received FAA approval to burn auto fuel in addition to avgas. Auto fuel can look almost clear when viewed in the right conditions. At least one auto fuel manufacturer now produces clear fuel, making it even more difficult to tell what has gone into your airplane's fuel tanks. If you are unsure about what is in the fuel tanks, find out. It might take having the FBO check its fuel records, but it is worth the effort. Having the wrong fuel can reduce engine performance, increase engine wear, or cause the engine to quit completely.

After completing the fuel inspection, I examine the left side of the fuselage, looking for ripples or dents in the skin that might indicate a structural problem. If all looks satisfactory, I move on to the left wing inspection.

Preflight: Left wing. The first thing I do is look down the leading edge of the wing. As with the left side of the fuselage, I am interested in deformations of the leading edge that could be a sign of damage or structural problems. If I find no problems, I move under the wing, beginning at the wing root. I verify that the fairings, which in the case of the Tomahawk are rubber gaskets, are properly secured. While under the wing, I look at the underside of the fuselage for any rivet damage or skin deformation. Following this I move on to the left main gear, looking for proper tire inflation and wear. I check to see that bolts, nuts, and cotter keys are in place, and I check brake wear and brake line condition. I also look at the ground around the tire for any brake fluid stains.

If the aircraft you are flying has wheel pants, it might be more difficult to see each of these items, but try to verify each of them as best you can. Next I drain fuel from the left wing tank into the drainer, checking it for the items discussed during the left cowl inspection. I then inspect the ground around the drain, looking for fuel stains that might indicate a leak in the fuel system. The pitot tube is then inspected for any obstructions, and I also give the stall warning vane a slight upward tap to make sure it is not stuck. If the aircraft is tied down, I then remove the left wing tie-down. As I perform these inspections under the wing, I look around for any loose or missing rivets or skin damage.

At this point I move above the wing, removing the fuel cap to check the level of fuel and its color. Remember, 80 octane is red, 100 octane is green, and 100LL is blue. If you have a color other than the one specified for your aircraft, you have a problem that needs further investigation before you fly. We have all heard stories about general-aviation aircraft that have had jet fuel mistakenly pumped into the tanks. One hundred feet in the air is not the time to find out that reciprocating engines don't run well on JP4. After securing the fuel cap, I continue to move down the wing to the tip, looking for skin damage and missing, loose, or sheared rivets. At the wingtip I look along the top and underside of the wing, hunting for any signs of structural problems, and then I look over the wingtip for any missing screws or cracks. I also inspect the position light, strobe lights, and any other objects fastened to the tip.

The ailerons are next on the inspection list. I move them through their full range of motion, feeling for any binding or sticking. Following this I look at each of the piano hinges that attach the aileron to the wing, making sure they are securely attached and that there is not excessive play between the halves of the hinge. Additionally, I look at the turnbuckle and make certain it is securely attached to the aileron and pushrod.

Now I come to the flaps, which were previously lowered. I like to make sure the turnbuckle and pushrod are secure and the flap hinges are not worn. I give the flaps a general once-over to make sure they are not damaged or corroded. After the flaps I continue down the left side of the fuselage and look for popped rivets and bent or buckled skin until reaching the tail.

Preflight: Tail surfaces. The Tomahawk is a T-tail configuration, so it is a little harder to properly inspect than the tail surfaces normally found on general-aviation trainers. However, a satisfactory preflight of this area can still be accomplished. I start at the bottom of the rudder, inspecting the pivot and rudder cable attach points and looking for any missing bolts or cabling that are not properly attached. I then look up the rudder and verify that bolts are in the rudder/vertical stabilizer attach points. At times it is necessary to feel in areas that I cannot see to make sure the bolts are there. I then inspect the left elevator/horizontal stabilizer, checking for any missing screws in the tip fairing. As I move along the elevator from the left to the right side, I make sure the attach bolts are present. These are easy to see and pose no challenge for the inspection. Finally, I check the right elevator tip in the same manner as the left. If the tail has a tie-down attachment, I remove it at this time.

Preflight: The right side of the aircraft. I continue down the right side of the fuselage and wing, inspecting them in the same manner that I described for the left. When completed, I am back at the starting point, the right-side cowling. After reading this you might be asking yourself, "Is all this really necessary? I don't have half an hour to do a preflight every time I fly." The answer is yes, it is necessary. But it doesn't take more than a few minutes to do once you get the routine down.

Early in my flight training I did preflight inspections like many pilots. They were relatively quick walk-arounds, during which I'd give the airplane a general once-over. I felt this was the same airplane I flew every week and did not really expect to find anything wrong with it. On one flight the instructor asked me if I had completed the preflight and was ready to go. Of course I was. The instructor walked to the tail of the aircraft and removed the rudder lock that I had missed. This was something new the FBO had put on the aircraft and I hadn't bothered to look up to see it. This oversight left a strong impression on me. What else was I missing by thinking there was never anything to see on a preflight? It was then that I started to look and poke and prod during preflight inspections, and I eventually developed the routine that was just described. This preflight takes about five minutes and is worth the extra effort involved. When I finish, I am certain as I can be that the aircraft has no problems and is ready to fly.

If you are flying at night, you can still do a preflight as thoroughly as during daylight. But make sure you have a good flashlight with strong batteries in it. I recently did some night flying, and about halfway through the preflight the light started getting dimmer. I was able to finish, but I had to put the flashlight closer to everything to inspect it. One other point about the night preflight: check your navigation lights to be sure they are all working.

I recently took a friend flying for the first time with me. This individual has been around flying for some time and has flown as a passenger in general-aviation aircraft with a number of pilots. During the preflight he commented he has never seen one done as thoroughly. You don't have to be an aircraft mechanic (A&P) to do the preflight correctly. Look for items like those described in the preflight example; oil leaks, popped rivets, and deformed skin are just a few examples. You should also review your aircraft's operations manual for any additional preflight details like acceptable engine oil levels or where to check the fuel system for water.

The key to a good preflight is be observant. You don't have to know what is causing the pool of oil in the bottom of the cowl, but when you see it you will recognize that something is wrong, and you can have a mechanic look at it before you fly.

WEATHER

The final topic in this chapter is weather. Within the scope of this book it is impossible to give in-depth coverage of all aspects of weather and how it can affect flight. The intent of this section is to discuss some of the more crucial weather factors that we need to be concerned with during takeoffs and landings. To that end we will review wind shear and temperature and humidity.

Wind shear

Wind shear is a sudden, strong variation in wind direction and, possibly, wind velocity, between horizontal layers. Figure 1-1 is a simple representation of wind shear. At 2000-feet AGL the wind's velocity is 20 knots. At 1500 feet, it has changed direction 180°, and the velocity is 15 knots. Finally, at 1000 feet it has increased velocity to 25 knots and has again changed direction. Wind speed does not affect the aircraft's airspeed. However, it does have a direct impact on ground speed and flight path. Let's briefly review what private-pilot training covered concerning the effects of wind.

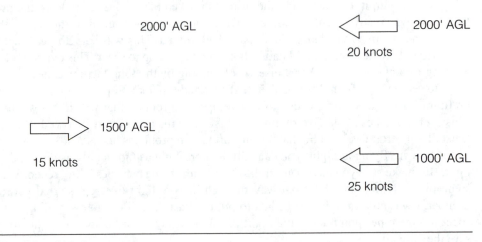

Fig. 1-1. *Wind shear.*

As we fly, the airspeed indicator uses the difference in air pressure between the pitot tube and static air vent to generate the indicated airspeed it displays. We then take the indicated airspeed and correct it for temperature and pressure to arrive at the true airspeed. This is the value we use when computing the effects of wind on ground speed. If we are flying at a true airspeed of 100 knots and there is no wind, our speed over the ground is also 100 knots. If we are flying at the same 100 knots into a direct 20-knot headwind, we can subtract that 20 knots from our true airspeed to arrive at a ground speed of 80 knots. A 20-knot wind blowing from directly behind the airplane would be added to the 100-knot true airspeed for a ground speed of 120 knots. The change in wind direction does not cause our indicated airspeed to change, but it causes significant changes in ground speed.

Let's put this in the context of a landing. Figure 1-2 displays an aircraft on final approach. As you can see, the airspeed remains constant, but the ground speed varies as the aircraft descends through each layer of the wind shear. Assume a constant rate of descent and no changes to correct for variations in ground speed. Figure 1-2 also represents the effect of wind shear on the glidepath. Beginning at position 1, the pilot has

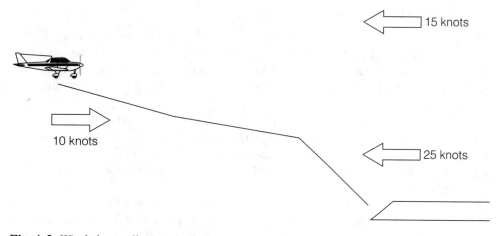

Fig. 1-2. *Wind shear effect on ground speed and glideslope.*

set up a glidepath that will take him to the runway under normal conditions. At position 2, the wind shear has become a tailwind, and the glidepath has flattened out. The aircraft is still descending at the same rate, but the higher ground speed now causes it to cover more forward distance in a given period of time, causing the glideslope to have a shallower angle. The airplane is flying into a stronger headwind at position 3 and is now covering less distance over the ground in a given period of time. The constant rate of descent causes the glideslope to steepen. As the approach progresses down through each layer of wind shear, the glidepath is affected. The cumulative effect is a landing short of the runway.

Today many tower-controlled airports have wind-shear detectors placed around the perimeter and are able to sense when wind shear is present. Controllers can then warn arriving and departing aircraft. I was recently doing touch and goes at a local tower-controlled field during wind shear conditions. The wind shear was not a headwind/tailwind switch of direction as in the previous example, but a side-to-side shift. At about a quarter-mile final, I could feel the airplane being rapidly pushed from one side, then the other, as I descended down through the layers of the shear. Fortunately, the tower was broadcasting wind shear alerts and I was able to compensate, but for me it was a very real example of how wind shear affects the airplane.

At uncontrolled fields, or controlled fields with no sensors, you need to rely on yourself to detect the effects of wind shear. If your flight path changes as a result of wind shear, compensate by adjusting power settings as required to control glidepath, or adjust for directional changes the shear might cause. In extreme situations, it is better to go around and try again than to attempt to salvage a bad situation. In the next section we will discuss two other weather-related topics: temperature and humidity.

Temperature/humidity

Changes in weather can affect your aircraft's performance. In this section we discuss how temperature and humidity cause these performance variations. In chapter 10 we review performance charts. The following topics directly relate to those charts as they depict changes in takeoff and landing distances as a result of different temperature and humidity levels.

Air density is directly affected by the temperature of the air. This same air density affects how much horsepower is produced by the engine, how much thrust is produced by the propeller, and how much lift is produced by the wings. Another factor related to air density is humidity, or the relative content of water vapor in the air. As the relative humidity of air goes up, its density goes down.

Pilots who have taken off on a hot, humid summer day know that they used more runway than in the middle of winter. The reason is the difference in the density of the air. On a hot day, the air molecules have more energy as a result of heating. Any gas that is heated expands in volume, and as the volume it occupies becomes larger the molecules are spread over a greater distance. This results in a reduction of density within a given area. Hot-air balloons are a graphic example of this principle. The warmer air within the balloon is less dense than the surrounding air and is buoyed up, causing the balloon to rise. Because it is less dense, warm air has less oxygen in a given volume for the airplane's engine to burn. As a result, less horsepower is produced.

When air cools, it has the opposite effect on air density. The molecules have less energy, moving around less and allowing more to fit in a given area. Now the air has become denser. This allows the engine to produce more horsepower because more oxygen is available to burn in a given volume.

Humidity, or water vapor in the air, has a similar effect on air density for a different reason. Water vapor molecules are lighter than air molecules, and water-vapor molecules reduce the density of the air in a given volume because they displace the heavier air molecules. The higher the water vapor content, or relative humidity, the less dense the air.

This higher-density altitude, which is the actual altitude corrected for nonstandard temperature and pressure, results in higher true airspeeds and a need for longer runways. It is generally felt that high humidity has a smaller effect on density altitude than temperature and barometric pressure.

An aircraft engine produces horsepower by burning a fuel/air mixture. As we climb in altitude, we lean the mixture because there is less oxygen to burn in a given volume of air, so less fuel is needed to keep the mixture at the correct level. The higher an airplane climbs, the less power the engine produces. In a similar manner, the reduced volume of oxygen in hot, humid air also reduces the power the engine can produce. The result is a longer run for the aircraft to achieve takeoff speed because less power is available from the engine to produce thrust.

The propeller is basically an airfoil moving through the air. When air is not as dense due to high heat and humidity, it is not able to produce as much thrust. This, combined with lower power from the engine, is why high heat and humidity can make what seemed like a long runway last winter too short during the summer. It pays to understand how heat and humidity affect an aircraft's performance. When you are flying from paved, 10,000-foot runways, you might not give temperature and humidity a second thought other than how much it makes you sweat while you are doing the preflight.

But on shorter runways, knowing the effect they have, combined with using the aircraft performance charts, could make the difference between a safe takeoff and ending up in a corn field at the end of the runway.

SUMMARY

We have covered several different areas that are directly related to takeoffs and landings in this chapter. First we discussed the fact that how often and what type of flying you do is directly related to how well you take off and land. The more often you practice, the more you will improve the consistency and level of your performance. Experience is the best teacher, and the only way to get it is by doing. Make a commitment to yourself to improve how well you fly, and then go practice.

The second topic of chapter 1 dealt with preflights. Many pilots today "kick the tires and light the fires" and feel that is enough. It's not. You owe it to yourself and your passengers to do a thorough preflight inspection. As was already stated, it takes only a few minutes to do a good preflight and can make a big difference in the outcome of your flight. What if a bird's nest inside your cowling caused you to overheat the engine, doing several thousand dollars worth of damage? This is just one example of why you need to pay attention to detail during your preflight.

The final section in the chapter dealt with weather factors that you need to be particularly concerned with during takeoffs and landings. Temperature and humidity have a tremendous effect on how an aircraft performs, and you need to be aware of their impact when you fly. Wind shear is another weather factor that can cause problems if you aren't paying attention. By reacting quickly enough to wind shear, you can avoid trouble.

As you continue to read keep the topics discussed in this chapter in mind. They are a foundation to build on and are related to what is discussed later.

2
V-speeds

ONE OF THE MAJOR KEYS TO WELL-EXECUTED TAKEOFFS AND LANDINGS IS airspeed control. Knowing what airspeed to use is also crucial. In this chapter we will discuss several airspeeds, or V-speeds, that you should be familiar with. There are a fairly large number of V-speeds, each with a defined meaning. We will review a small subset of those airspeeds, giving the definition of each one. Where to find the values associated with each V-speed will also be covered, in addition to FAA-published recommendations for airspeeds to use during an approach. Finally, because they are so closely related to airspeed control, stall avoidance and recovery are examined.

V-SPEEDS

V-speeds provide a standardized method for defining airspeeds. In this section, seven V-speeds will be reviewed. This will include not only a definition of the V-speed, but also a discussion of the rationale behind it.

V_{SO} (stall speed landing configuration)

First on the list of speeds we are going to review is V_{SO}. V_{SO} is the airspeed at which an airplane stalls in landing configuration. It can vary for each aircraft, but landing configuration is generally considered to be full flaps and power off at maximum landing weight. For retractable-gear aircraft, this also includes gear down. You will occasionally hear this configuration referred to as "dirty" because so much drag is being generated by the extended gear and flaps. Drag is also produced by the aircraft itself due to its high angle of attack.

V_{SO} is a critical airspeed. During an approach to land, it is necessary to remain above this airspeed to avoid a stall. Yet you must maintain an airspeed relatively close to V_{SO} to allow a smooth transition from flare to touchdown. Excessive speed carried during the flare can result in the aircraft floating down the runway until surplus airspeed has bled off. The aircraft could also balloon, or rise back into the air (Fig. 2-1). If this is not corrected, the aircraft can stall and drop to the runway at too high a rate of descent. This could result in anything from rattled teeth to a damaged aircraft, or worse. Too little airspeed, and the pilot could end up in a low-altitude stall, from which recovery can be difficult, if not impossible. In chapter 4, we will review recovery from unusual landing situations such as ballooning.

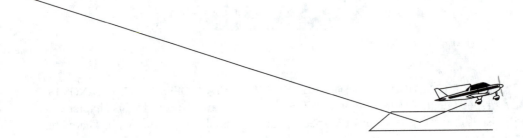

Fig. 2-1. *Aircraft ballooning.*

V_S (stall speed clean configuration)

V_S is the stall speed in a clean configuration. This means power off and flaps retracted, as well as gear up for retractable-gear aircraft. V_S is higher than V_{SO} because lift-enhancement devices, such as flaps and slats, are not used to help generate greater lift at slower airspeeds.

Remember that both V_{SO} and V_S are stalling speeds in level flight. As you increase the angle of bank during a turn, the stall speed also increases. (See Fig. 2-2). In a 60° bank, the airplane's stall speed is approximately 1.4 times the stall speed in straight and level flight. More to the point, if your Cessna 172 has a V_S of 44-knots indicated airspeed (KIAS), that value will increase to almost 62 KIAS in a 60° bank. It's easy to see how low, slow, steep turns onto final can become a dangerous situation.

Stall recovery

This is a good place to review stall recovery. The best way to handle an approach or departure stall is to never get into one. Always pay attention to your airspeed and attitude.

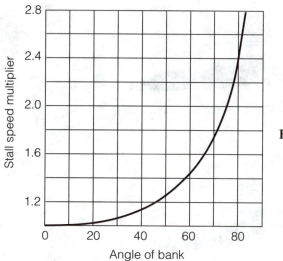

Fig. 2-2. *Bank/stall speed.*

Keep your eyes moving as they scan the panel and outside the aircraft. Think ahead and plan what you need to do as you fly the airplane so that you don't get behind the aircraft. Falling behind often leads to fixation on one item or another, further aggravating problems. Don't set yourself up for trouble by letting the airspeed get too slow or the angle of bank too steep.

The first sign of an impending stall is stall buffet. This is caused by the turbulent flow of air over the wings as the critical *angle of attack* is approached. The angle of attack is the angle formed between the cord line of the wing and the relative wind (Fig. 2-3). The *cord line* is an imaginary line extending from the trailing edge of the wing through the center of the leading edge. *Relative wind* is the flow of air that is parallel, and opposite to, the aircraft's flight path. Recall that the critical angle of attack is the angle of attack at which airflow over the wings becomes so turbulent it no longer produces the necessary lift. Also remember that the critical angle of attack can be achieved at any airspeed and flight attitude. Don't get into the habit of thinking that because you are flying at speeds well above V_S or V_{SO}, you can't get into a stall situation.

If you find yourself in a low-altitude stall, the first rule is to maintain control of the airplane. Lower the nose slightly by reducing back pressure on the yoke or stick, and apply full throttle. Control any rolling tendencies by using rudder, not the ailerons. Use of ailerons for roll control can aggravate the stall situation.

As soon as you have attained flying speed, resume level flight or enter a slight climb. Avoid the tendency to try to climb too quickly. This could result in the entry into another stall. The goal is to avoid ground impact and stall reentry. You will need to decide the best course of action based on your altitude and any obstacles in your vicinity.

Low angle of attack

Medium angle of attack

Excessive angle of attack. Air flow is
broken. No lift to
sustain the airplane.

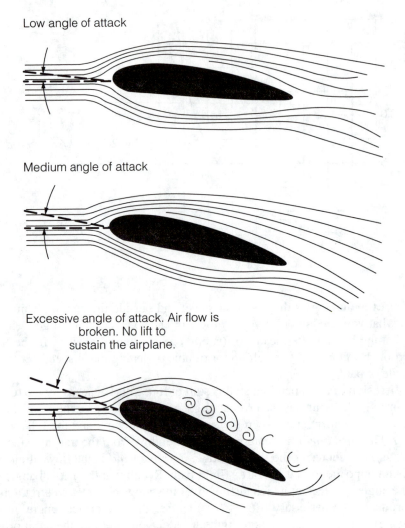

Fig. 2-3. *Critical angle of attack.*

V_A (maneuvering speed)

Maneuvering speed, also known as V_A, is the airspeed at or below which sudden, full-control movements or turbulence cannot cause structural damage to an aircraft at gross weight. V_A is not normally marked on the airspeed indicator. If it is not placarded on the instrument panel, you should look it up in the aircraft's operations manual.

The maneuvering speed is an important number to remember. If you encounter heavy turbulence, reduce your airspeed to at or below V_A. At airspeeds above V_A, turbulence itself can cause damage to the aircraft due to the sudden G-loads imparted to the aircraft. The same is true of sudden control movements. If you review the opera-

16

tions manual for your aircraft, you will likely find that as the gross weight of the plane changes, so does the V_A. Be sure to use the correct V_A for the weight you are flying at.

V_{NE} (never-exceed speed)

V_{NE}, or the never-exceed speed, is the maximum airspeed at which you should fly an aircraft. At speeds above V_{NE}, structural damage can be inflicted on the airplane due entirely to the speed at which it is traveling.

As with V_A, excessive control movements or turbulence above V_{NE} can subject the aircraft to undue G-forces, causing structural damage. In some cases control flutter also can be induced by exceeding V_{NE}. *Control flutter* is the rapid movement or oscillation of control surfaces. In severe cases this fluttering can shake an aircraft apart within seconds of its onset. The important message here is never exceed V_{NE}; it could become a situation that gets out of control very quickly.

V_{FE} (maximum flap-extended speed)

V_{FE} refers to the maximum airspeed at which flaps should be in the extended position. Above this airspeed, the flaps, pushrods, flap guides and other flap-related hardware can be damaged. The damage is caused by loads placed on them by the excessive airspeed. Damage can range from bending or twisting flaps to completely losing them. The range of airspeeds at which flaps can be used is explained in greater detail in a later section covering airspeed indicator markings.

V_X (best angle of climb)

The best angle of climb speed allows you to gain the most altitude in a given distance. This is often the airspeed recommended by aircraft manufacturers for climbing after a short-field takeoff. A short-field situation allows you to gain the most altitude in the distance between the lift-off point and any obstacles in your path.

V_X should not be flown at for extended periods of time. To achieve V_X the nose of the aircraft must be relatively high. Climbing at this higher angle of attack reduces the flow of cooling air into the cowling, and after a period of time causes higher engine operating temperatures. With repeated, prolonged climbs at V_X, the life of the engine can be reduced, resulting in the need for an early engine overhaul.

V_Y (best rate of climb)

To achieve the greatest altitude gain in a given period of time, use the best rate of climb airspeed. Aircraft manufacturers often recommend V_Y as the cruise climb airspeed. V_Y is flown at a lower angle of attack than V_X. This allows more air to flow through the engine's cowling to cool the engine, lowering the engine's operating temperature. This, in turn, reduces wear during cross-country climbs.

If you happen to fly an airplane with a cylinder-head temperature gauge, you can verify the difference in temperatures between V_X and V_Y. This same effect can also be measured with the oil temperature gauge. In the Cessna 182 I fly, I have observed a 30° to 40° difference in oil temperatures between flying at V_X and V_Y on hot, summer days. Aircraft and engine manufacturers have spent a great deal of time and money to arrive at the V_X and V_Y speeds they recommend, so it is best to take advantage of the results of their testing and experience.

THE AIRSPEED INDICATOR

Now that we have covered a few V-speeds, let's discuss the airspeed indicator. Several of the airspeeds we have covered in this chapter are marked on the airspeed indicator. Additionally, several airspeed ranges are also represented on its face.

Figure 2-4 displays a typical airspeed indicator. The first airspeed marking we will review is red line. This corresponds to V_{NE}, which was previously defined. This mark represents the maximum airspeed the aircraft can be flown at. Leading up to red line is a yellow arc. This arc is used to caution pilots that they are approaching red line.

Fig. 2-4. *Airspeed indicator.*

The green arc represents a safer range of airspeeds to operate the aircraft at. Colocated with a portion of the green arc is the white arc. This arc indicates the safe airspeeds at which flaps can be used. The upper end of the white arc represents V_{FE}. Flaps should not be extended when the airspeed indicator is above the white arc. The lower end of the white arc corresponds to V_{SO}.

Most aircraft today have the arcs and red line printed directly on the airspeed dial. Some older aircraft have them painted on the glass face of the airspeed indicator. If you happen to fly an airplane with these types of markings, be certain the glass has not shifted in the case. If it has shifted, the airspeeds you think fall within a given range could be wrong due to the rotation of the glass.

Another factor related to the airspeed indicator must also be considered. The FAA states that the calibrated airspeed should be used when computing the correct approach speeds for a given landing weight. Many aircraft with manufacture dates prior to the mid-1970s have airspeed indicators with speed range arcs marked in calibrated airspeed (CAS) values. But for many aircraft manufactured after the mid-1970s the values marked are in indicated airspeeds. Airspeed indicators marked with indicated airspeed require a conversion to compute the correct calibrated airspeed. Figure 2-5 is an example of an airspeed correction table for a fictitious aircraft. You can see that indicated airspeed (IAS) can be substantially different from calibrated airspeed at slower speeds, and this becomes more pronounced as the aircraft approaches stall. For example, with an indicated airspeed of 40, the calibrated airspeed with flaps up is actually 53. With flaps down, the same indicated airspeed becomes 49 CAS. Indicated airspeed is incorrect due to placement of the pitot tube and airflow into it. As the aircraft's angle of attack changes with slower airspeeds, the effect of the error becomes greater. Tables are provided by each aircraft manufacturer and can be different for each model aircraft, so be sure to use the correct version for your aircraft.

IAS	40	50	60	70	80	90	100	110	120	130	140
Flaps up	CAS	53	59	63	73	78	88	98	106	118	128	136
Flaps down	CAS	49	56	62	72	82	91	100	*	*	*	*

Fig. 2-5. *Airspeed correction table.*

It is clear that only a few of the airspeeds we need to know are marked on the airspeed indicator. In some cases, additional airspeeds, such as V_A, are placarded on the panel. However, many are still not readily available. For those airspeeds, you need to look in the aircraft's operations manual. I like to make up a table of airspeeds for each type of aircraft I fly and attach it to my kneeboard for reference before and during flight. I have also seen several different versions of "flip cards" for sale that serve the same purpose. Basically they have fill-in-the-blank areas that allow you to document the airspeeds for easier review. You should know, or have ready access to, V-speeds when you fly.

V-SPEEDS IN THE PATTERN

The FAA has set recommended speeds that general-aviation aircraft should use in the pattern. These speeds are multiples of V_{SO} and should be used as guidelines. Before we get to the FAA's recommendations, a word of caution: if your aircraft's operations manual specifies airspeeds other than the FAA recommendations, use the manufacturer's speeds.

To begin, the FAA rule-of-thumb states that on the downwind leg of the pattern, use airspeeds that are not greater than V_{FE}, or less than 1.4 times V_{SO}. The FAA also recommends that the airspeed should be no lower than 1.4 times V_{SO} until on short final, at which time you can reduce it to 1.3 times V_{SO}. If high winds, turbulence or icing conditions are encountered during approach, adjust your airspeed upwards as necessary to maintain safety margins ("On Landings, Part I," p. 3).

Let's apply the recommended airspeeds to a real world example. We will use a Cessna 172 with a V_{SO} of 33 KIAS and V_{FE} of 85 KIAS as an example. On entering downwind you could use an airspeed between 85 and 47 KIAS. These are the high (V_{FE}) and low (1.4 times V_{SO}) airspeeds, according to the guidelines. You would use airspeeds within this range (for instance, 80 KIAS on downwind, 70 on base, and 60 on final) until on short final, at which time you would reduce airspeed to at or above 43 KIAS. (Caution: remember to correct KIAS to get CAS). During flare and landing, airspeed would be reduced to allow for a controlled touchdown.

SUMMARY

In this chapter we have reviewed a number of different topics associated with airspeed. Initially we defined several V-speeds that are of concern during takeoff, approach, and landing. Among the V-speeds were V_{SO} (the stall speed in the landing configuration), V_{FE} (the maximum flap-extended speed), V_X (the best angle of climb speed), and V_Y (the best rate of climb).

Stall recovery techniques were also examined. The important points to remember are maintain control of the aircraft and execute stall recovery with a minimum of altitude loss. The best method is to avoid stall situations by being "situationally aware" while in the takeoff and landing phases of your flight.

Airspeed indicator markings were also discussed, with an explanation of each arc's meaning. Finally, the FAA's recommended speeds for use during approach and landing were introduced. The primary concept that you should commit to memory is that airspeed maintenance is extremely important. Controlling and maintaining sufficient airspeed affords you safe control of the aircraft. Remember to adjust your airspeed as necessary to allow for any wind and weather factors. By using the information covered in this chapter, you will fly within a safe range of airspeeds for your aircraft. Knowing and applying this knowledge will make for safer, more accurate takeoffs and landings.

3
The traffic pattern

FLYING IN THE VICINITY OF AN AIRPORT DEMANDS THAT PILOTS DIVIDE their attention among many different areas. Pilots are flying relatively low over the ground, and they are either cleaning up the aircraft's configuration after takeoff or preparing it for landing. They are also monitoring the airspeed, engine instruments, and flight instruments. While performing all of these tasks, pilots must also monitor and talk on the radio, scan for traffic, and plan the entry or exit from the airport traffic pattern. Flying safely in this environment requires planning and preparation. This chapter discusses a number of aspects associated with approaching, flying in, and departing from an airport traffic pattern. Among them are radio use, markings used to indicate traffic-pattern direction, working at tower-controlled fields, pattern entries and exits, aircraft separation, and busy-pattern considerations. By planning ahead, pilots can reduce the work load associated with pattern flying and make it easier to stay in control of aircraft.

INITIAL RADIO CALL APPROACHES

Whether you are approaching an uncontrolled or controlled field, you should begin listening to the appropriate frequency well away from the airport itself. At an uncontrolled field, an initial call should be made while you are 5 to 10 miles out, so it is appropriate to begin listening five or more minutes before that initial call. At controlled airports, it might be necessary to make an initial call even further out, depending on the type of airspace you are flying into. For example, at many type-C airspace airports, you are required to make the initial call to approach control between 15 and 20 miles from the airport. It is important to monitor the appropriate automated terminal information service (ATIS), approach, and tower frequencies well before initial radio contact. During this monitoring phase, you should be generating a mental picture of traffic flow at the airport. At controlled airports, this will give you an idea of how controllers are positioning aircraft, allowing you to anticipate how you will be placed into the flow. At uncontrolled fields, monitoring the common traffic advisory frequency (CTAF), also known as unicom, can help you plan your entry into the pattern with minimum disruption to other traffic. At this point we will discuss radio communications at uncontrolled and controlled fields. Let's begin with uncontrolled airports.

All radio communications should be brief and to the point. At uncontrolled fields, the same CTAF might be used by a number of different airports in the same geographic area, creating a contention problem for pilots attempting to use that frequency. This manifests itself with frequent squeals as multiple pilots attempt to transmit at the same time. Before making a radio call, listen for other radio traffic. The initial call to the airport should state the airport being called, the aircraft's make, registration number, location, altitude, and a request for the winds and active runway. After the call, wait to receive an answer before repeating the request. At uncontrolled fields, the line personnel are frequently engaged in other activities and might take a few seconds before they can respond. If no response is received after several requests, listen for other traffic at the airport. Aircraft in the pattern should state what runway they are using, giving you the active runway information.

As you approach the airport pattern, state your intentions before entering it. For example, if you plan to enter the pattern on a crosswind or downwind leg, broadcast that plan so other aircraft are aware of your position and what you are going to do. I also like to state my position during the approach as I turn onto base and final. If you do this while in the turn, it will give other aircraft behind you in the pattern an easier target to spot. Any unusual activities, such as executing a touch and go, should also be stated in advance. After landing and leaving the runway, it is good practice to radio that you are clear of the active. It might be necessary to eliminate some of the previously mentioned calls after initial pattern entry if the pattern and radio traffic are very busy. In those cases it is best to state your position in the pattern as radio traffic allows.

At this point we will move on to tower-controlled airports. Because of the diversity of communication facilities at controlled airports, it is not possible to discuss all of them. We will, however, cover several major areas of interest. Among these are ATIS, approach control, the tower, unicom, and CTAF. The automated terminal information service is a valuable aid to pilots. It is a continuous recording of information broadcast on the ATIS frequency. This information is updated as often as changing conditions require, and it contains information about current weather, communications, runways, approaches in use, and other pertinent data the pilot might need. Each revision is identified by a phonic letter of the alphabet such as alpha, beta, etc. An example of an ATIS broadcast is:

"INFORMATION GULF. MADISON INFORMATION FOR 1753 ZULU. CEILING 25000 BROKEN, VISIBILITY 15. TEMPERATURE 18. DEWPOINT 3. WIND IS 330 DEGREES AT 8. ALTIMETER SETTING 30.55. AIRCRAFT APPROACHING FROM THE WEST CONTACT APPROACH ON 124.0, FROM THE EAST ON 120.1. DEPARTING AIRCRAFT CONTACT CLEARANCE DELIVERY ON 121.62. AIRCRAFT ARE TAKING OFF AND LANDING ON INTERSECTING RUNWAYS. INFORM ON INITIAL CONTACT THAT YOU HAVE INFORMATION GULF."

When ATIS is available at a facility, pilots should listen to it prior to making the initial radio call. When the initial call is made, the pilot should state the facility being called, the aircraft make and identification number, the location, altitude, intention to land, and the fact that the pilot has the ATIS information. For example:

"MADISON APPROACH, THIS IS PIPER ONE SEVEN SIX ONE DELTA, TEN MILES TO THE SOUTHWEST, 5000 FEET, LANDING, WITH INFORMATION GULF."

When the controllers know a pilot has the information available from the ATIS broadcast, they will not repeat it to the pilot. This reduces the work load of the controller and the amount of radio traffic.

Always use the aircraft's full identification number during the initial call. On any subsequent transmissions, you should state the message or request along with the aircraft identification in one brief statement. Unless otherwise instructed, acknowledge instructions or frequency changes you are given by approach or the tower. This becomes even more important if there are aircraft in the pattern with similar identification numbers. If an emergency requires you to deviate from the instructions you are given, notify the tower as soon as possible after the deviation.

Sectional charts and the Airport/Facility Directory list the communication services available at an airport. If there is a unicom frequency for a tower-controlled airport, it will always be 122.95. This can be used to communicate needs for fuel and other ground-related services. When a tower does not operate continuously, the CTAF should be used when it is closed. The CTAF frequency in this situation will always be the tower frequency. Pilots should use the CTAF during periods when the tower is closed, in the same manner as at an uncontrolled field.

DEPARTURES

When departing an uncontrolled field, the CTAF should be used to communicate your intentions to other aircraft. This can include taxiing or backtaxiing on a runway, crossing a runway, and departing. As during the approach, listen to find out what other traffic in the airport's vicinity is doing prior to your departure. At controlled fields, you should listen to ATIS prior to the initial call. Depending on the services available at the controlled field, you might need to contact clearance delivery to receive a departure clearance, ground control to receive taxi clearance, tower to receive takeoff clearance, and departure after takeoff and while still in the airport's airspace. Consult the appropriate section chart, Airport/Facility Directory, and other documentation to find the correct procedures and frequencies for each airport you visit.

LEFT AND RIGHT PATTERNS

At most uncontrolled and controlled airports, the traffic pattern consists of a series of left turns. The basic pattern is made up of an upwind leg, a crosswind leg, a downwind leg, a base leg, and the final approach leg (Fig. 3-1). When flying in the pattern, or at any time, the pilot should be scanning the airspace for other traffic. Under visual flight rules (VFR), it is the pilot's responsibility to watch for and avoid other aircraft. The majority of aircraft are designed with the left seat for the pilot-in-command, making it easier for the pilot to scan airspace to the left rather than the right. As you fly a left pattern, it will be easier to see other aircraft in your vicinity. However, for reasons such as obstructions, simultaneous use of runways, and noise abatement procedures, patterns might be set up with turns to the right.

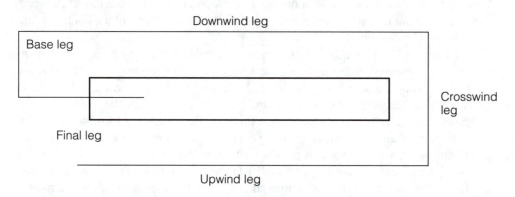

Fig. 3-1. *Pattern legs.*

You can determine whether an airport uses left or right patterns from several sources. The first is the Airport/Facility Directory that is published by the U.S. government. This document contains a great deal of information about airports. Among this data is when an airport has a runway that uses right traffic patterns. Figure 3-2 is a listing for the Fond Du Lac County airport from the EAST CENTRAL U.S. Airport/Facility Directory. As you can see, the entry for runway 18 documents that right traffic is in use on that runway. Also note in the remarks section that runway 36 uses only left turns in the traffic pattern (Airport/Facility 1994, p. 203). If you do not have access to an Airport/Facility Directory, you can contact a flight service station (FSS) and ask FSS personnel if right patterns are in use during your briefing. Most fixed-based operators (FBOs) also have copies of the Airport/Facility Directory on hand that you should be able to use.

FOND DU LAC CO (FLD) 1 W UTC–6(–5DT) N43°46.25'W88°29.31'   CHICAGO
   807   B   S4   **FUEL** 100LL, JET A1   TPA—1807(1000)   H–36, L–12E
   RWY 18–36: H5560X75 (ASPH)   S–22   MIRL   IAP
      RWY 18: REIL. Trees. Rgt tfc.   **RWY 36:** REIL. VASI(V4L)—GA 3.0° TCH 35'.Road.
   RWY 09–27: H3602X75 (ASPH)   S–22   MIRL
      RWY 09: Trees. Rgt tfc.   **RWY 27:** Road.
   AIRPORT REMARKS: Attended 1400Z‡–dusk. For arpt attendance after dusk, phone 414-922-6000. Rwy
      36 is for left tfc only. MIRL Rwy 18–36 preset on low ints; to ACTIVATE higher ints and MIRL Rwy
      09–27 and REIL Rwys 18 and 36—CTAF. VASI Rwy 36 ops 24 hours.
   COMMUNICATIONS: CTAF/UNICOM 122.8
      GREEN BAY FSS (GRB) TF 1-800-W-BRIEF. NOTAM FILE GRB.
      RCO 122.5 (GREEN BAY FSS)
   ® **CHICAGO CENTER APP/DEP CON** 127.0
   RADIO AIDS TO NAVIGATION: NOTAM FILE OSH.
      OSHKOSH (L) VORTAC 111.8   OSH   Chan 55   N43°59.43'W88°33.36'   165° 13.5 NM to fld. 780/2E.
      NDB (MHW) 248   FLD   N43°46.17'W88°29.08' at fld. NOTAM FILE GRB. Out of svc indefinitely.
      SDF 108.3   FLD   Rwy 36

Fig. 3-2. *Airport/Facility Directory listing.*

Another source of information might be the airport itself. As you approach, markers that are visible from the air might be laid out on the ground (Fig. 3-3). These indicate the direction of turns in the pattern. The L-shaped markers in the figure indicate that right turns are used. If you are uncertain about pattern turn directions, stay above pattern altitude until you are able to determine the correct direction. It is then recommended that you remain clear of the pattern while descending to pattern altitude.

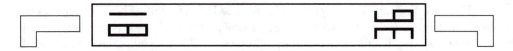

Fig. 3-3. *Pattern direction indicators.*

When flying a right pattern, you should be especially vigilant during turns. In high-wing aircraft, the lowered wing will have a tendency to block your visibility in the direction you are turning. In a descending turn, a low-wing aircraft might block your view of the airspace beneath you. Be sure to clear your airspace before you begin a turn. Every year, aircraft collide due to pilots taking for granted that no other planes are in their airspace.

As you fly a right pattern, your perspective in relation to the runway will change. For instance, you might use the runway's relationship to a wing strut to approximate the correct distance you want to achieve during the downwind leg. In a right pattern, this will change and you might need to use a different reference point on the right strut to achieve the same distance from the runway. While making turns onto base and final, you might also find that the wing or cabin structure blocks your view of the runway. To compensate, it might be necessary to lean forward to look around the wing or other obstruction to see the runway. As you do this, don't forget about your airspeed and aircraft's attitude. Pilots unaccustomed to leaning forward during a right turn to base or final might inadvertently pull or push the control yoke as they turn. As a result, the aircraft might gain or lose airspeed, resulting in potential stalls or ground impact if the situation is not recognized and corrected in time. You might also find it is more difficult to judge the correct radius of a right turn to final with a tailwind on the base leg. If you are uncomfortable with flying right patterns, contact your favorite flight instructor and find an uncontrolled field with right patterns or a controlled field and request right patterns from the tower. It will be practice well-spent.

GROUND MARKINGS

As you approach an airport for landing, there might be a number of ground references that give you information that is useful for landing. Among these are the wind sock, the wind tetrahedron, and the segmented circle. Runway markings are also useful in determining the type of runway and where it is safe to take off or land on the runway surface.

Wind sock

The wind sock (Fig. 3-4A) is probably the most common wind indicator in use at airports. It is frequently a bright orange color and mounted on top of a building or pole near the runway. It generally has a large opening on the end attached to the pivot point and a smaller opening at the far end from the pivot point. As a result of this design, it will swing the smaller, trailing edge away from the wind, indicating the direction the wind is from. You can also gauge the relative speed of the wind by how straight out the sock is extended. If the wind sock is hanging down, the wind speed is relatively low. If it is extended slightly, there is a light to mild wind. If it is fully extended, wind speeds are higher.

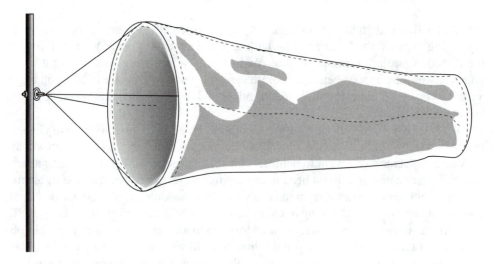

Fig. 3-4A. *Wind sock.*

Wind tetrahedron

The wind tetrahedron (Fig. 3-4B) shows the direction the wind is from by turning into the wind. Its triangular shape causes the pointed end to turn into the wind, pointing in the direction aircraft should land. These are usually located near the runway or runway intersection, and at night they might be lighted. They do not indicate relative wind speeds, as the wind sock does. You should also be aware that in light or no-wind conditions the tetrahedron might not point in the correct direction of the wind, due to its own weight being too heavy for the wind to turn.

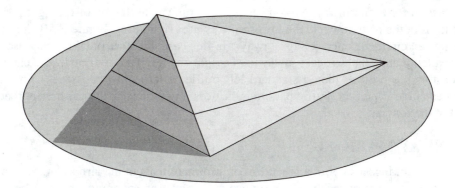

Fig. 3-4B. *Wind tetrahedron.*

Landing tee

The landing tee (Fig. 3-4C) is similar in function to the tetrahedron. It is shaped like an aircraft and pivots on its mount. Like the tetrahedron, it will point into the wind to indicate the direction aircraft should land. As with the tetrahedron, it will not indicate

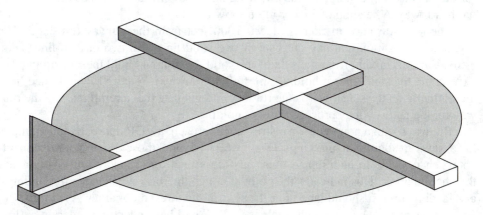

Fig. 3-4C. *Landing tee.*

29

relative speed of the wind, and in light or no-wind conditions it might not indicate the correct direction of the wind. It might also be lighted to allow for use at night, and it is normally located near the runway edge or intersection of runways.

Segmented circle

The segmented circle acts as a centralized point for wind direction and traffic pattern information. Figure 3-5A displays a segmented circle located between parallel runways. The wind tetrahedron in its center indicates the landing direction. Traffic pattern indicators located around the perimeter of the circle indicate the direction of turns in the pattern for each runway. As you can see, the runway at the top of the figure uses left turns in the pattern, while the lower runway uses right turns. Figure 3-5B shows a segmented circle for intersecting runways. In this case, the horizontal runway uses a left pattern, while the vertical runway uses a right and left pattern. If no traffic pattern indicators are present, then a standard left traffic pattern must be used. If no segmented circle is present, traffic pattern indicators might be located near the approach end of the runway.

RUNWAY MARKINGS

Runway markings also offer a great deal of information about the airport and runway. From the air, the pilot can tell if the runway has an instrument approach, distances on the runway, portions of the runway that are not suitable for landing, and a number of other items. Figure 3-6A is an example of a visual runway's markings. This runway does not have an instrument approach to it. Markings on it consist of the centerline markings, the designation markings, a fixed-distance marking (if the runway is longer than 4000 feet and used by jets), and holding-position markings.

The designation marking indicates the approach direction of the runway. The direction is the whole number value closest to the one-tenth value of magnetic direction for the runway. For example, a runway heading of 60° will be shown as the number 6 on the runway. A heading of 240° will be shown as 24.

The fixed-distance marker is placed 1000 feet from the runway threshold. The taxi-holding-position marking consists of two solid lines and two dashed lines (Fig. 3-6B). Aircraft taxiing from the side of the solid lines should hold their position prior to reaching the solid lines if holding short of the runway is required. The dashed lines are normally on the side towards the runway and indicate that aircraft leaving the runway for the taxiway may proceed without holding.

Runways with a nonprecision instrument approach, a VOR approach for example, have several additional markings (Fig. 3-6B) beyond those for a noninstrument-approach runway. In addition to those just described, the runway will also have a threshold marker. This indicates the point at which the runway is usable for landings, and landing aircraft should land after the start of the threshold marker. A runway with a precision-instrument approach, such as an ILS, will have even more markings (Fig.

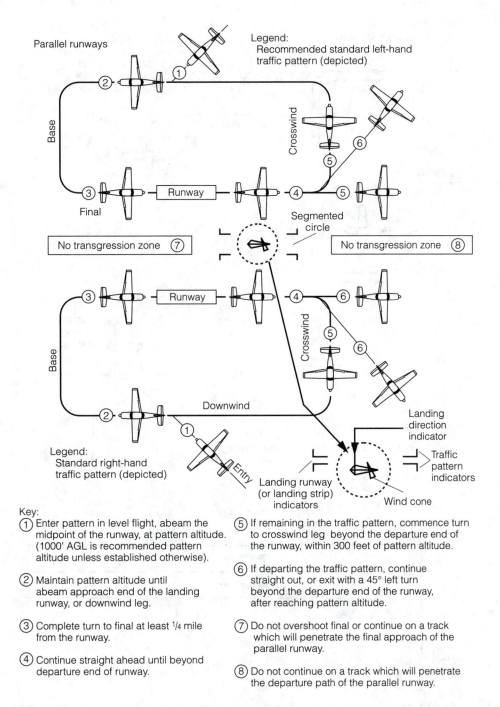

Fig. 3-5A. *Segmented circle/parallel runways.*

Key:

① Enter pattern in level flight, abeam the midpoint of the runway, at pattern altitude. (1000' AGL is recommended pattern altitude unless established otherwise).

② Maintain pattern altitude until abeam approach end of the landing runway, or downwind leg.

③ Complete turn to final at least 1/4 mile from the runway.

④ Continue straight ahead until beyond departure end of runway.

⑤ If remaining in the traffic pattern, commence turn to crosswind leg beyond the departure end of the runway, within 300 feet of pattern altitude.

⑥ If departing the traffic pattern, continue straight out, or exit with a 45° left turn beyond the departure end of the runway, after reaching pattern altitude.

⑦ Do not overshoot final or continue on a track which will penetrate the final approach of the parallel runway.

⑧ Do not continue on a track which will penetrate the departure path of the parallel runway.

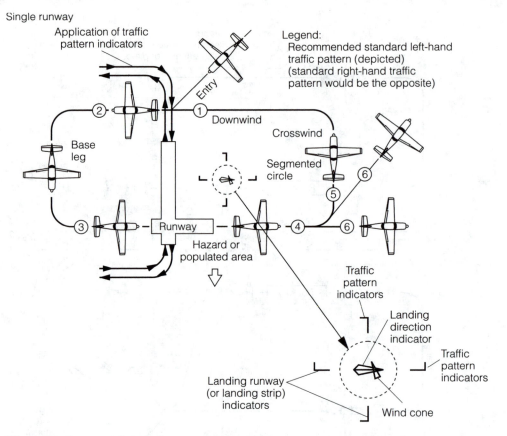

Fig. 3-5B. *Segmented circle/intersecting runways.*

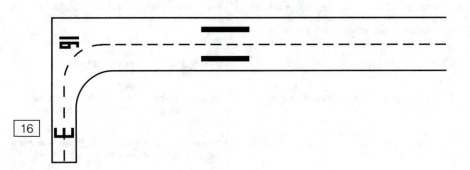

Fig. 3-6A. *Runway markings/simple.*

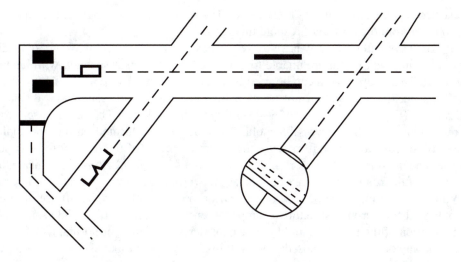

Fig. 3-6B. *Taxiway hold lines.*

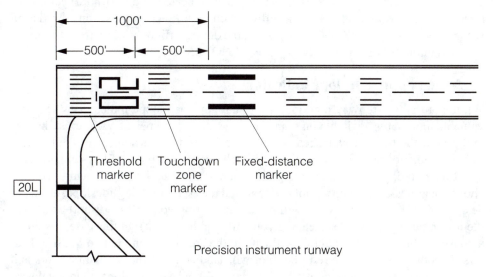

Fig. 3-6C. *Precision-instrument runway markings.*

3-6C). These can consist of the threshold marker, the designation marker, the touch-down zone marker, the fixed-distance marker, and side stripes. The touchdown zone marker is located 500 feet from the runway threshold. The fixed-distance marker is located 500 feet beyond the touchdown zone marker, or 1000 feet from the threshold marker. Side stripes consist of four sets of stripes that begin past the fixed-distance marker. The first two sets consist of two parallel lines on each side of the runway. The last two sets are single lines on each side. The start of the last single side stripes are 3000 feet from the runway-threshold marker.

Runways will frequently have overrun areas that will support the weight of an aircraft taxiing over it, but are unable to withstand the stresses associated with landing aircraft. This overrun area can be used by aircraft taking off or completing a landing rollout, but the area should not be used for touchdown while landing. A displaced threshold (Fig. 3-7) is indicated by several types of markings. The first is a set of arrows pointing to the runway threshold. These arrows are located on the nonlanding area of the runway. Aircraft may use the arrowed portion of the runway for taxi purposes and the start of a takeoff roll. The second is a series of chevrons that indicate that portion of the runway may only be used as an overrun or stopway for landing aircraft. While this portion of the runway might appear to be suitable for aircraft use, it might not have the underlying structure to support even a taxiing aircraft's weight safely, and this portion of the runway should not be considered available for normal use. Finally, if a runway or taxiway is closed to aircraft use, large Xs will be visible on it. Aircraft should not use these runways or taxiways. I have seen some temporary runway closures at night marked by large, lighted Xs placed at each end of the runway. Runway closures will be included in NOTAMs and are also located in the Airport/Facility Directory if the runway is closed for a sufficient duration, such as grass runways during winter. The closure NOTAMs are also available during a flight briefing from flight service, assuming they have not become class II NOTAMs.

Visual approach slope indicator

While not really a runway marking, the visual approach slope indicator (VASI) is a valuable tool that is available on many runways. Several types of VASI exist, but we will review the two-bar white/red type. Figure 3-8 is a representation of the possible light combinations a pilot can see on approach with a two-bar VASI system. When a VASI bar is red, you are below the glideslope for that bar. When it is white, you are above the glideslope. When a pilot has achieved the correct VASI glideslope, normally designed for 3°, the upper VASI will be red (indicating you will land short of it), and the lower bar will be white (meaning you will land beyond it). In this case, assuming a constant glideslope, the aircraft will land in the area between the two VASI indicators. If both VASI bars are red, the pilot will land short of the VASI indicators and should shallow the glideslope until intercepting the VASI glideslope, then readjust it to maintain the correct glideslope. If both are white, the pilot will land beyond the VASI indicators, and the pilot should steepen the glideslope until the correct glideslope is intercepted. At this point the pilot should again readjust to maintain the correct glideslope.

PATTERN ENTRY

Earlier in this chapter we used Fig. 3-1 to represent the airport pattern. Now let's review each leg of the airport pattern in more depth. To begin, traffic pattern altitude is

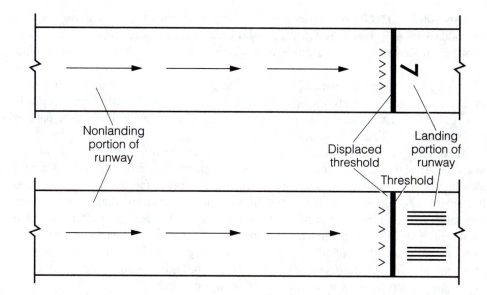

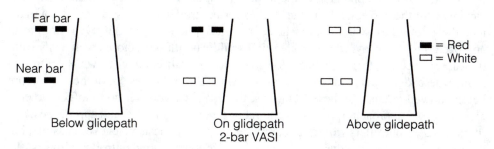

Stopway and blast pad area

Fig. 3-7. *Displaced thresholds.*

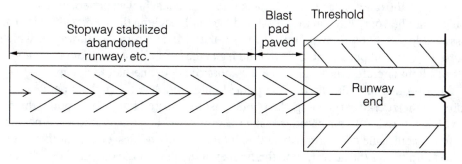

Fig. 3-8. *VASI lighting.*

normally 1000-feet AGL. A number of airports have pattern altitudes that might range from 600 to 1500 feet, so it is always a good idea to consult the Airport/Facility Directory for the correct pattern altitude when flying into an airport you are unfamiliar with.

The upwind leg of the pattern begins at the point the aircraft becomes airborne. For aircraft already in the air, the upwind leg of the pattern begins when the aircraft is flying parallel to the runway centerline used for takeoff. The upwind leg of the pattern continues until the aircraft makes a 90° turn.

There are several schools of thought on when the turn to crosswind should be made. For aircraft taking off, some references indicate that when the aircraft is 300 feet below pattern altitude, you should make the turn. Some pilots turn to the crosswind leg after achieving 500-feet AGL during the climb, while others turn when they feel they have achieved an altitude that would permit a safe return to the airport after an engine failure (more about this topic in a later chapter). I prefer to make the turn to crosswind after achieving an altitude of 500-feet AGL, continuing to climb as I do. The following factors might influence the decision of when to turn: tall hills, radio towers, aircraft capability, or ground congestion such as tall buildings or antennas.

The 90° turn from upwind takes the aircraft to the crosswind leg. This leg is perpendicular to the runway centerline. If the aircraft has not achieved pattern altitude, you should continue your climb on the crosswind leg until reaching the correct altitude. The crosswind leg should be flown until you reach the distance from the runway, at which point a 90° turn will put you on a downwind leg at the correct distance from the runway. For most light aircraft, this distance will be approximately one-half to one mile.

After turning onto downwind, the pilot should complete the prelanding checklist to prepare the aircraft for landing. At a point past the desired touchdown point, the pilot will then turn 90° to the base leg. Like the crosswind leg, the aircraft is perpendicular to the centerline of the runway and will be slowing and descending at this point. The last turn brings the aircraft onto the final approach leg of the pattern. Here the aircraft is aligned with the centerline of the runway, where it continues its descent to landing. The preceding explanation assumes no-wind conditions and that the aircraft will not need to compensate for wind. We will discuss pattern wind correction later in this chapter.

Aircraft will be approaching the airport from all angles, affecting the most efficient way for them to enter the pattern. At tower-controlled airports, controllers will direct the pilots during the approach, so at times pilots might be given straight-in approaches to a runway, while at other times pilots might be vectored well away from the airport as controllers sequence pilots into existing traffic. I have flown at controlled fields where, due to large air carrier traffic, I was vectored miles out of the way before being allowed to land. In these situations, there is little you can do to affect the efficiency of your approach. It is possible to request various runways and approaches, and in most cases controllers will try to comply with your requests whenever possible, but they will still vector you during the approach. However, at uncontrolled airports, the situation is much different.

At uncontrolled airports, you are responsible for planning your approach. This includes tasks such as deciding when to let down to pattern altitude, what angle to approach the pattern from, how to sequence into the pattern, and what other traffic is in the pattern. Among the many activities you have as you fly, it cannot be emphasized often enough that it is your responsibility to constantly scan for other traffic whenever you fly. Each year there are midair collisions, both in ATC-controlled and uncontrolled airspace because pilots become complacent about scanning for other traffic. It is always your responsibility to look for other aircraft when you fly.

You should also be aware of the types of other aircraft in the pattern. If the aircraft in front of you is significantly slower than you will be during the approach, you will need to anticipate flying longer legs in the pattern to maintain sufficient spacing from it. If the aircraft behind you is a higher-performance aircraft, you might want to try to keep your speed higher during the initial approach phase, slowing more as you reach short final. This type of planning and "situational awareness" make it easier for you and others to fly safe approaches. When other aircraft are ahead of me in the pattern, I will normally stay on downwind until they pass me going the opposite direction on final. At that point I will normally turn to base and continue my approach. This helps assure that proper spacing will be maintained and that I will not be approaching the other aircraft in what could turn out to be a contention for the same airspace.

As you approach an uncontrolled airport, it is best to approach toward the center of the airport. From that point it is possible to enter the pattern from a number of different positions. We will now review some possible pattern entry positions. Keep in mind as you review the following material that not all situations can be discussed. Factors such as the amount of traffic in the pattern, obstructions, and local regulations might change what the correct pattern entry method might be for minimal traffic disruption and sufficient spacing. What might be safe in one situation might compromise safety in another, and it is up to you to determine the safest course of action.

Crosswind pattern entry

One method of entry is the crosswind pattern entry. Position A in Fig. 3-9 represents how an aircraft approaching the runway from the opposite side of the pattern can enter at midfield over the runway on a crosswind entry. After crossing the active runway at a perpendicular angle, the aircraft continues until in the correct position to turn from the crosswind to the downwind leg. This basic crosswind entry can be used when traffic is sufficiently light that you will not be cutting off someone else in the pattern as you make the entry. When traffic is higher volume and aircraft spacing is closer, it might not be safe to enter the pattern in the manner just described. In these situations it might be necessary to change the crosswind entry from midfield to further back in the pattern, or to extend it out far enough that essentially it becomes a turn to a longer downwind approach leg. You will need to determine if the conditions permit safe entry into the pattern from a crosswind approach, and plan accordingly.

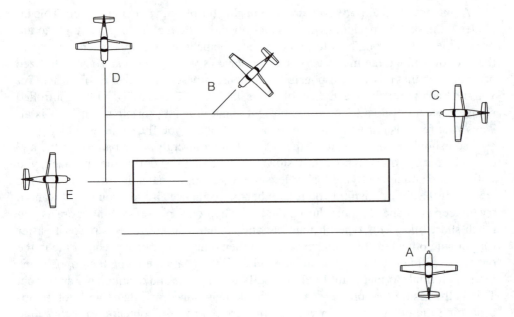

Fig. 3-9. *Pattern entry points.*

Downwind entry

The most common type of pattern entry, and the one recommended as the safest by most FAA-published references I have reviewed, is the downwind pattern entry. Position B in Fig. 3-9 depicts the classic 45° downwind pattern entry, where the aircraft enters the downwind leg at a 45° angle at pattern altitude. As the aircraft is approaching the pattern, the pilot is in a position to easily scan for other traffic already in the pattern. In much the same manner that a car merges onto an interstate highway from an on-ramp, the pilot merges into the flow of traffic in the pattern. The 45° angle of intercept also allows the pilot to turn to the correct downwind heading without an excessive bank angle.

While in the pattern, it is a good idea not to make turns with banks in excess of 30°. In chapter 10, stall speeds will be discussed in more detail, but remember that as your bank angle increases, the stall speed also increases. A common reason for aircraft accidents in the vicinity of airports is that pilots don't plan their turns properly, and they use very steep banks, at low airspeed, trying to correct their mistake. These low, slow, steep turns cause the aircraft to stall, without sufficient altitude to recover. Keep your banks less than 30°, and if the situation begins to get out of control, go around and try again.

Another variation is the straight approach to downwind entry. An aircraft might approach the airport from a direction and on a heading that allows it to fly straight into the downwind leg of the pattern. Figure 3-9, position C, shows how the aircraft would approach the pattern and enter it. In this case, the pilot must again be aware of traffic and allow for adequate spacing during sequencing.

Base entry

When an aircraft enters the pattern on a base leg (Fig. 3-9, position D), it flies directly onto base and bypasses the downwind leg of the pattern. If done safely, and in the correct traffic conditions, this can save time and fuel. However, there are a number of difficulties associated with this type of pattern entry. First is maintaining sequencing with other traffic in the pattern. I have heard more than one comment made on unicom to the pilot who cuts in front of other aircraft already in the pattern by flying directly onto the base leg. From a safety standpoint, it is more difficult to determine spacing from other aircraft and the impact of your aircraft on traffic flow. It might pose problems of seeing other aircraft in the pattern and being seen by them. If you enter on a base leg, you will very likely be at a lower altitude than aircraft on downwind, and they might lose you in the ground clutter.

There is also a problem with determining the correct altitude and airspeed to enter the base leg on. When you start from downwind, you have set up your descent, airspeed, and power from a known point, and the results are fairly predictable. If you misjudge your altitude, power setting, or airspeed when entering from the base leg, it becomes more difficult to compensate as you try to correct the situation. In other words, you have added additional factors to your approach that complicate your flying the pattern correctly. This is not an impossible situation, but it can cause additional work, especially for low-time pilots, or pilots flying into unfamiliar airports. My recommendation is not to use this type of approach at uncontrolled fields unless you are sure of the traffic, the airport, the airplane, and the weather (not to mention the pilot you might have cut off in the pattern wanting to exchange a few pleasantries with you). You are more likely to use the base approach at a controlled field. The controller will tell you to enter a base leg to a runway, and you should plan your approach accordingly.

Final entry

Entry to the pattern from final approach, also known as a straight-in approach, is illustrated in Fig. 3-9, position E. In this approach technique, the aircraft bypasses all other legs of the pattern and flies directly into a landing. Like the base-leg approach, the straight-in approach saves time and fuel when used. But also like the base-leg approach, there are similar problems with using it. Again, it is very possible to cut off other airplanes in the pattern. Also like the base leg, it is more difficult to see and be seen by other aircraft, for much the same reasons. With the straight-in approach, it is more difficult to judge your airspeed, altitude, power, and descent. If you are unsure of traffic, or any other factor, it would be better to enter an upwind leg, fly the crosswind leg, and enter on downwind. I do not recommend the use of the straight-in approach at uncontrolled fields. However, at controlled fields, it is very possible you could be set up by tower personnel for a straight-in approach to final.

Every airplane will fly differently, so it is difficult to describe the correct way to fly a straight-in approach that will work in all situations. At approximately one to two miles from the approach end of the runway, the airplane should be slowed to approach speed and have an altitude from 500 to 800 feet AGL. At one-half-mile final, the aircraft should be in its landing configuration, and the rate of descent should be at the appropriate value. At this point, the altitude, airspeed, and power settings would be the same as though the approach had been started from downwind. Be sure to listen on the radio, scan for other aircraft in the pattern, and announce your position frequently during your approach. Also be sure that radio towers, high power lines, and other obstructions, which are not a factor in normal approaches, do not present a hazard during the straight-in approach.

PATTERN EXIT

Now that we have reviewed the different pattern entries, we will look at pattern exits. How you exit the pattern can have an impact on other traffic and flight safety. At controlled fields, the tower and departure personnel will give you headings and altitudes to fly. But at uncontrolled fields, you must make the decision about what is the safest, most effective method for leaving the vicinity of the airport. If a higher-performance aircraft is taking off behind you at a controlled-field ATC, if possible, you will be vectored out of the flight path of the other aircraft to allow them to take off and not overrun you. However, at an uncontrolled field, this is something you as the pilot need to be aware of and make your own flight path adjustments. As always, remember to scan for other aircraft as you fly.

Crosswind exit

Position A in Fig. 3-10 shows the path of a crosswind exit. In this situation, the aircraft climbs out after takeoff, and, after reaching the appropriate altitude, makes a 90° turn. The pilot continues to climb on this heading until clear of the airport traffic pattern. At this point the pilot turns to the desired heading and continues the flight. When using this method, you will follow the normal flow of traffic in the pattern and present less disruption. The altitude for the turn to crosswind has been previously discussed. One of the negatives of this type of pattern departure is that crosswind heading might be up to 180° from your intended direction of flight, wasting time and fuel as you fly out of the pattern and make the turn to the correct heading. While on crosswind, it is also more difficult to see aircraft approaching from the right side of the aircraft as they enter a left downwind. You will need to frequently scan for other aircraft during the crosswind exit.

A variation of the crosswind exit is the 45° exit. Figure 3-10, position B, indicates how an aircraft, after achieving a safe altitude, can make a 45° turn from the runway

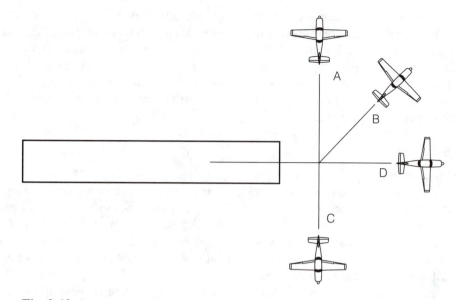

Fig. 3-10. *Pattern exit points.*

heading to exit the pattern. There are some benefits in using this type of exit. Your initial turn can be with a shallower bank, allowing you to continue the climb with less loss of lift during the turn. This type of departure also makes it easier for you to see aircraft that might be on or entering a downwind leg. In this situation, the aircraft are more likely to be located somewhere off the nose of your airplane, as opposed to off the right wing. This makes it easier for you to scan during the departure. (But don't neglect to look to the right!) Using the 45° departure might make it easier to turn to your desired heading, reducing wasted time.

Right-turn exit

The right-turn pattern departure is depicted in Fig. 3-10, position C. In this situation you exit the pattern, a left pattern for this example, by turning to the right. This type of departure is used by pilots when they do not want to waste time turning left to exit the pattern, then right to turn onto heading. Whatever the justification, most pilots have exited the pattern with a right turn from time to time. We have previously stated one apparent advantage, but there are also disadvantages. During the right turn it might be more difficult to see aircraft entering the pattern on a crosswind or upwind leg. Going against the standard flow of the pattern, other aircraft might not be as likely to be looking for you coming from an odd direction. The question you must ask yourself as you plan this type of departure is, "Is this type of departure safe under the current situation?" You might want to consider the number and type of other aircraft in the pattern. Also, listen to the radio to determine how arriving aircraft are approaching the pattern.

If there seems to be a potential conflict with arriving airplanes, you might not want to make this type of exit. As you make your decision, keep in mind that safety is the ultimate goal, and saving a few minutes of flying time might not outweigh the risk. Like the base- and final-approach pattern entries, I do not recommend the use of this exit for a left pattern. You will most likely see this departure at a controlled field at the request of pilots who want a right-turn departure after takeoff or when controllers need it to be used for traffic separation reasons.

Straight-out exit

The last pattern exit we will cover is the straight-out departure (Fig. 3-10, position D). In this case you might have an on-course heading that approximates the runway heading. As you depart you will continue your climb and runway heading while you leave the airport vicinity. One consideration you should have is the type of airplanes that might be taking off behind you. As was previously discussed, if a higher-performance aircraft is waiting to depart behind you, it might have to delay its takeoff or make additional turns to avoid airspace contention with you. Also remember to transition to the appropriate climb speed recommended by the aircraft manufacturer. Changing the normal sequence of steps during departure could cause you to forget to transition from V_X to V_Y, for example, as you continue to fly straight ahead.

AIRCRAFT SEPARATION

We have already discussed the need for scanning while flying. But what is the correct distance to maintain from other aircraft while in the vicinity of an airport? What factors should be considered in determining how much distance should be between you and the aircraft in front of you? In this section we will analyze some considerations you should keep in mind as you decide what the safest distance is to follow other aircraft.

At a controlled airport, tower personnel will set the distance between you and other aircraft in the pattern. They will vary the distance based on the size and speed of aircraft. For example, if you are following a large, commercial jet, there might be a three-mile distance between you and the jet in front of you. Recently the FAA has been giving consideration to increasing the minimum required spacing behind a large airliner to five miles. The reason for this potential change is wake turbulence, which will be discussed in more depth later.

The problem for controllers comes when there is a pattern full of airplanes with differences in landing speeds. Even single-engine general-aviation aircraft can have sufficient differences to cause a spacing problem. I was making a landing at a tower-controlled airport with parallel runways. The aircraft I was flying was a Piper Arrow, and the airplane in front of me was a Cessna 150. On final I began getting closer to the slower trainer, and I started to do small S-turns to maintain my distance. As the Cessna approached short final, I was doing large S-turns to keep from running over it. During one S-turn that took me almost into the final approach course of the other parallel runway, the tower saw there was no longer sufficient distance between me and the Cessna and cleared me to land on the other runway. We had started out too close in the pattern, and as the Cessna slowed, there was no room to maintain adequate distance at a safe flying speed. When I noticed this, I should have notified the tower rather than wait for the situation to deteriorate.

Wake turbulence

When lift is generated by the wings of an airplane, there is a lower-pressure area located above the wing and a higher-pressure area below it. At the wingtip, there is a tendency for the higher pressure air to "roll" over the wingtip to the lower-pressure area above it. This rolling air generates a vortex of air that trails behind and below the aircraft (Fig. 3-11). Vortexes have a sink rate of approximately 500 feet per minute and have a tendency to level off about 900 feet below the aircraft generating them.

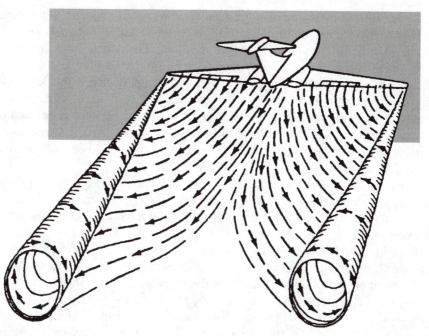

Fig. 3-11. *Wingtip vortexes.*

This vortex, also called *wake turbulence*, is strongest when an aircraft is heavy, clean, and slow. As a result, when aircraft are taking off and landing, they are generating the greatest amount of wake turbulence. Large, heavy aircraft will generate a stronger vortex than smaller, lighter planes. This becomes of special interest to the light-aircraft pilot because the wake turbulence from a large aircraft can be strong enough to cause your airplane to roll. There are many instances where general-aviation aircraft have crashed due to loss of control after being caught in the rotating forces of the wake turbulence behind a large aircraft.

To avoid a potential problem, it is necessary to take one of several courses of action. By maintaining sufficient distance behind a large aircraft in the pattern, you will allow time for the wake turbulence to dissipate. As previously mentioned, wake turbulence has a tendency to descend below the flight path of an airplane. Wind will also blow the vortex in its direction. All of this contributes to the movement of the vortex away from your flight path if given sufficient time and the right wind conditions. However, in calm conditions, vortexes generated by aircraft close to the ground have a tendency to move laterally across the ground away from the aircraft that generated them. A light, quartering tailwind can create the most hazardous situation due to the vortex being blown back towards the runway and potential landing zone. There have been cases where small aircraft have encountered wake turbulence several minutes after a larger aircraft has taken off or landed.

If, due to traffic density or other factors, you cannot maintain the distance you would like, your flight path should remain above the flight path of the landing aircraft, and you should land past the point where the other plane touched down. After touchdown, the landing aircraft's wings quit producing lift and, consequently, wake turbulence. If you are taking off after a large aircraft has departed, you should lift off prior to its rotation point, climb above its flight path, and stay upwind of its flight path. You should also be aware of the vortexes generated by aircraft on intersecting and parallel runways and avoid the flight paths of those aircraft as well. Wind can generate an appreciable amount of movement of the vortexes, and you should consider them as you plan your flight path.

In certain situations, tower controllers warn pilots of the potential for encountering wake turbulence. However, it is always the pilot's responsibility to adjust the operation and flight path of the aircraft to avoid wake turbulence. This is true whether or not a warning has been received.

Keep in mind that the roll rate of a vortex can very easily exceed the roll rate for many small, general-aviation aircraft. In these cases, full aileron opposite the vortex's direction will not be sufficient to maintain the aircraft's attitude. There have been cases where pilots find themselves inverted and try to do a half loop from the inverted position to regain control. Very often there is too much altitude lost, and the ground impedes any further progress. Given the opportunity, it is best to continue the roll back to the wings' level flight. If you have never performed a roll and would like to become more comfortable with this type of maneuver, find an aerobatic flight instructor and learn some of the basic aerobatic maneuvers. For those not interested in becoming competition-level aerobatic pilots, most aerobatic schools offer basic courses that cover spins, rolls, and loops. Knowing the feelings associated with being upside down in an airplane, or in any other unusual attitude, it is an immense confidence builder to know the correct control inputs to right the situation with a minimum of altitude loss. This is not to say you should go looking for a wake-turbulence-induced roll, but by being familiar with flying in unusual attitudes, you will be better suited to cope with them.

Visual spacing

We have discussed spacing while flying at airports with operating control towers and the effects of flying behind large aircraft. Our next topic will be safe aircraft spacing in the pattern at uncontrolled fields. Without a controller it becomes each pilot's responsibility to maintain adequate spacing. This requires constant scanning and thought as you smoothly transition from entering the pattern to landing and keeping sufficient space. You must be aware of the type of the other aircraft in front of you and plan your approach while anticipating theirs. If you are flying a high-performance single and the airplane in front of you is a slow trainer, you will want to start further behind the other airplane while still on downwind. As it begins to land, you might find that you need to extend your downwind leg to give yourself adequate spacing. As previously mentioned, you should not turn on to base, unless you are flying a very large pattern, until the other aircraft has passed you while inbound on final. If you are flying a lower-performance aircraft, you will not have to allow for as much spacing if the plane in front of you has similar or higher-speed landing characteristics.

Small, general-aviation aircraft do not generate wake turbulence with the same force as larger aircraft. It is created, but not to the extent that you should remain several miles behind other light aircraft to avoid it. The FAA recommends that light aircraft maintain a minimum of 3000 feet lateral spacing while in the pattern. By adhering to this, you increase the reaction time you will have in the event aircraft in front of you have an emergency situation, execute an unexpected maneuver (i.e., a 360° turn on final), or are much slower during landing than originally anticipated.

Radio use

In addition to visually scanning for other aircraft to preserve proper separation, the radio serves as a useful tool. By broadcasting your intentions while in the pattern and listening for where other aircraft are, you can increase your level of understanding of the flow of traffic in the pattern. We have previously discussed the initial call prior to entering the pattern and the need to listen prior to the first call to generate a mental image of what the traffic will be when you arrive at the pattern. But what if you hear other aircraft stating their positions and you cannot see them? The best course of action is to contact them and find out if they have spotted you. It is also good practice to broadcast what aircraft you are following and that you have them in sight. This lets others in the pattern know more about sequencing of each aircraft. An example would be:

> "MADISON TRAFFIC, PIPER SEVEN NINER ONE DELTA ALPHA NUMBER THREE TO LAND BEHIND THE CESSNA 172 ON DOWNWIND."

This lets everyone know exactly where you fit into the flow of traffic and reduces any confusion others might have.

THE BUSY PATTERN

At one time or another, every pilot will encounter the situation of six or eight aircraft in the pattern when you want to either depart the airport or try to land there. In these cases it can be difficult to get onto the runway to take off and still maintain sufficient spacing or to try to fit into the pattern on a downwind approach to land. You will need to understand the situation and decide what is reasonable and safe under these circumstances, but we will cover a few of the criteria you should consider.

Pattern entry

It might seem difficult to enter a pattern that has a large number of aircraft in it and still maintain flight safety. In these cases you should consider what other aircraft are doing in the pattern and plan your entry to downwind with maximum spacing and minimum disruption in mind. You should already have a general idea of how many airplanes are in the pattern as you approach it. The safest approach will likely be the 45° entry we discussed earlier. This gives you a wide view of the pattern as you approach it and makes it easier to sequence into the downwind leg.

As you are approaching downwind from this position, consider the aircraft you will be in front of and behind. If you will end up between two higher-performance aircraft, you might want to keep your airspeed up to avoid spacing problems with the other aircraft, reducing your airspeed as you prepare for landing. If you are following slower aircraft, you should be prepared to reduce your airspeed to allow for adequate spacing.

I have seen situations where pilots have entered a busy pattern from other than the 45° entry. In one case a pilot entering from crosswind was cut off by another aircraft on downwind, requiring the pilot to exit the pattern and reenter from the 45° angle. When making the crosswind entry on busy days, it is more difficult to sequence into the pattern and can require larger changes in airspeed and banks. As you make the turn, the aircraft you are in front of will very likely be gaining on you, and it is also very possible you will lose sight of them during the turn to downwind. While it might cost you a little more time to set up for the 45° entry, it can make it easier to fly the approach. Every situation is unique, so be flexible in your plans.

Takeoffs

When it seems like airplanes are spaced every quarter mile on final, it can be difficult for you to get on the runway after an airplane has touched down. Wait for it to clear the runway and take off before the next landing airplane gets too close for comfort. I have had several situations where this was the case. There are some things you can do to improve the situation, though. First, be prepared to roll as soon as you turn onto the runway. Before taking the runway, make sure the flaps are set and everything else is ready. It wastes time and spacing to turn onto the runway and then finish setting the aircraft up for takeoff. Be prepared to roll onto the runway with power, and apply full power as soon as you are lined up with the centerline. This technique is used for soft-field takeoffs and can be used to save yourself a few seconds of time and give you some additional spacing from the next landing aircraft.

Consider the type of aircraft that are landing. If you are flying a typical two-seat trainer, it could be a problem to try to accelerate out of the way of a high-performance airplane. But if the mix of aircraft in the pattern includes other lower-performance planes, you might want to plan your departure while they are on final. The slower landing speeds they will have gives you more time to take off and keep adequate spacing. It might be necessary to act quickly to make a safe takeoff, so plan and be prepared to take advantage of an opening. Keep in mind that safety is always first and should not be compromised for any reason.

One thing that I do not recommend is beginning your takeoff roll prior to the last landing aircraft clearing the runway. I have seen this done, and it might seem like something you can get away with, but there are several risks. First, the aircraft might not make the turnoff you expect it to take. This would require it to continue to taxi down the runway. Second, I have seen aircraft have flats after they land, and in some cases the pilots had to slow down and taxi very slowly off the runway. At this point you might be telling yourself you can get off the ground before getting close to them, even in the two situations just described. But what if you have problems and have to abort the takeoff? It could cause problems you might not be able to avoid. Give it serious thought before ever trying this.

In the case of departing the airport vicinity, it might be best to use the 45° departure path that was previously discussed. This gives you maximum visibility while scanning for other aircraft in the pattern and reduces the amount of time you are in the flight path of other departing aircraft. You, of course, will need to judge the situation and determine what the best path is, but keep in mind how you will be affected by and affect others in the pattern.

PATTERN WIND CORRECTION

Private pilots learn ground reference maneuvers to help them compensate for the effects of wind while in the pattern. S-turns, turns about a point, and rectangular course all teach the student pilot what actions need to be taken to maintain a predetermined path across the ground, regardless of the direction of the wind. While it is not within the scope of this book to address this topic with the level of detail required for a student pilot, we will briefly review some key wind correction points.

You will want to fly a rectangular pattern with reference to the runway and path over the ground. On windy days this will require heading corrections into the wind to prevent you from drifting from the correct ground track. Figure 3-12 depicts a runway and ideal ground track for the pattern. As the pilot is flying the downwind leg, he or she will need to "crab" to the right to offset the effects of the wind pushing to the left. If the correct crab angle is not flown, the airplane will either drift towards or away from the runway. As the pilot turns onto the base leg, ground speed will increase. If this is

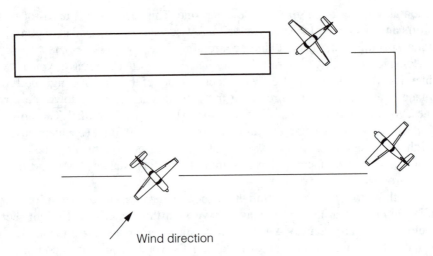

Wind direction

Fig. 3-12. *Crosswind pattern correction.*

not a direct tailwind, it would also be necessary to crab the aircraft to maintain the correct ground track. Because of the additional ground speed, it will be necessary to begin the turn to final earlier or use a steeper bank. Due to the crosswind from the left on final, the pilot will turn more than 90° in the turn to final to set up a crab for final approach.

Recently I flew on a day when the wind created a very strong headwind on base, and the Aerobat I was flying seemed to crawl toward final. Remember that strong winds can change the required glideslope and power required to touch down at a predetermined spot on the runway.

SUMMARY

We have covered a wide range of topics in the chapter. All of them relate to flying in the pattern near other aircraft and should be considered when you are in the vicinity of the airport. To safely fly in the pattern with maximum efficiency requires that you be "situationally aware" as you approach the airport and plan your landing. The same holds true while taking off. You should know what is going on in the pattern before you roll onto the runway.

I arrived to fly at a small airport I used to work at shortly after two airplanes had collided at the intersection of two runways. One airplane had been taking off and the other landing. No one survived the crash. Maybe each pilot thought the other should give right of way. Perhaps they were not even aware that another aircraft was around. The effect of the crash left a strong impression on me: we must be constantly vigilant whenever we fly. In situations like that, there are no second chances to go back and start over.

Use the radio to listen and broadcast; use your eyes to scan and your mind to plan. "Pilot error" all too frequently is the cause of aviation accidents, and you can increase the safety of your flights by knowing how to enter and exit the vicinity of the airport, what ground and runway markings are telling you, and where to find the information you need to know about an airport that is new to you. Fly the pattern consistently and follow the takeoff and landing procedures recommended by the aircraft manufacturer. There is no substitute for knowing your airplane and the correct procedures for flying it.

4
Landing techniques

EVERY GOOD LANDING COMES DOWN TO THE TECHNIQUE USED BY THE PILOT. This technique includes how the airspeed is controlled, power settings, picking the touchdown point, correct vision during the flare, the flare itself, and a number of other factors. This chapter will review the topics just mentioned, in addition to a number of others. These include flare-height judgment and differences between conventional and tricycle-gear aircraft during the flare and touchdown. We will also cover ground effect, unusual landing situations and approaches, groundloops, night landings, and wet runway considerations.

These topics become the foundation for the different types of landings we will discuss in later chapters. An understanding of the mechanics behind a good landing technique makes it easier for you to adapt to different airplanes, weather circumstances, and runway conditions.

LANDING CHECKLIST

In chapter 3 we discussed the need to divide your attention among many different activities as you approach the airport for landing. The more complicated the landing situation becomes, the more difficult it is to be sure that all steps in configuring the airplane for landing have been completed. This becomes even more true when flying a complex aircraft with a constant speed prop, retractable gear, cowl flaps, turbochargers, and a host of other systems. By forgetting to configure the aircraft correctly, you run the risk of having anything from an embarrassing moment to doing serious damage to the plane.

By using the landing checklist for your airplane, you can avoid the problem of relying on your memory to configure the aircraft and potentially missing an important step. Every year, seasoned pilots land with their gear up, a seemingly unforgettable step when landing. There is an old adage, "There are two types of retractable-gear pilots: those who have landed gear up and those who will." That statement is based on a certain amount of truth. If it is possible for you to forget to complete a step during landing configuration, then the odds are at some point in your flying career, you will.

I will never forget the day I was working on the ramp and turned around to see a high-performance, single-engine airplane starting to settle to the runway during landing with the gear still up. Just as the prop blades began to hit the runway, the pilot pulled back on the yoke, ballooned the airplane back up, lowered the gear, and completed the landing. The tips of the prop were curled back about six inches. I ended up removing the propeller and sending it off for repair. That was an expensive lesson for the pilot, but it was one I will never forget either.

The FAA recommends that when flying a retractable-gear aircraft, you have the gear down and locked at the same point on downwind each time you land. At the latest, you should have gear "down and locked" before reaching the approach end of the runway while on downwind. When you lower your gear, always verify that it is down and locked. Again, the FAA recommends that you touch the gear indicators and say out loud, "gear down" or "down and locked" or "three in the green." By doing this you don't automatically assume that the gear is down and not verify it. You should then again verify that gear is down after turning on final ("On Landings: Part III," p. 1–2).

Using the aircraft manufacturer's landing checklist reduces the chances of forgetting something important. For example, let's consider a sample checklist for two hypothetical aircraft. The first will be from a single-engine trainer, the second from a complex single-engine aircraft.

Trainer landing checklist

- ☑ Fuel selector: both
- ☑ Electric fuel pump: on
- ☑ Mixture: rich
- ☑ Check engine gauges
- ☑ Carburetor heat: on
- ☑ Airspeed: 60 to 70 KIAS
- ☑ Wing flaps: as required
- ☑ Seat belts: fastened

Complex single-landing checklist

- ☑ Fuel selector: both
- ☑ Electric fuel pump: on
- ☑ Mixture: rich
- ☑ Check engine gauges
- ☑ Carburetor heat: on
- ☑ Manifold pressure: set
- ☑ Propeller control: full RPM
- ☑ Airspeed: 60 to 70 KIAS
- ☑ Landing gear: down and three green indicator lights
- ☑ Wing flaps: as required
- ☑ Seat belts: fastened

As you can see, there are a number of common steps in both checklists. If you frequently fly different types of aircraft, it becomes easier to forget a step or two when not using a checklist, and you might still feel like you completed the important points. Many times, landing checklists are placarded on or near the instrument panel. This reduces the need to dig through the glove compartment or a side pocket to find the checklist.

You should be using the aircraft manufacturer's landing checklist for each aircraft you fly, but in addition there is an acronym, GUMP, that specifies key points that should also be checked again while you continue your approach. Table 4-1 serves as a general checklist to double-check several important items, but it can leave out some important steps for your aircraft, so do not rely solely on GUMP for landing. For instance, checking carburetor heat is not mentioned, nor are flap settings. The GUMP list doesn't begin to cover items like superchargers and air-conditioning units. The more complex or high performance an aircraft is, the more specialized the checklist for it. Nobody's memory is infallible, so it is in the best interest of flight safety to use the prelanding checklist provided by the manufacturer. Since the aircraft operations manual is one of the required documents for flight, this information should always be available when you fly.

Table 4-1. GUMP checklist

G	Gas	(Fuel on proper tank)
U	Undercarriage	(Landing gear down and locked)
M	Mixture	(Rich)
P	Prop	(Full RPM for constant-speed props)

AIRSPEED/POWER CONTROL

Student pilots frequently make the mistake of equating engine power settings with the airspeed they want to fly during an approach to landing. The throttle in a car controls the speed, so, to the student, it stands to reason the same is true for an airplane. This misconception needs to be corrected immediately.

Airspeed during landing is controlled by the pitch of the airplane, which the pilot controls with the control yoke. By pulling back on the yoke, or control stick, you will raise the nose of the aircraft. The result will be a reduction in airspeed. By pushing forward on the yoke, you lower the nose of the plane, causing an increase in airspeed. Understanding this concept is fundamental to good, consistent landings. You should be able to hold the airspeed you want during an approach within five knots above or below the required airspeed by maintaining a constant pitch angle during the approach. This airspeed control criterion is so important that it is a requirement for the private pilot practical test. When pilots frequently change the pitch attitude of an airplane, they are causing the airspeed to fluctuate. Constant airspeed control is one of the basic building blocks of good landings.

Engine power settings are used to control the glideslope of the airplane, not the airspeed. Being able to hold a constant glideslope is another major factor in high-quality landings. Knowing how to adjust your glideslope if it is not correct is also important.

If you are above the correct glideslope during your approach, you will need to reduce your power setting. This has the effect of increasing the rate of descent, allowing you to intercept the correct glideslope. As you intercept it, you will then need to increase your power to reduce the rate of descent and maintain the correct glide angle. A common error among student and low-time pilots is to dive for the runway if they are too high during approach, instead of reducing the power setting to increase the rate of descent. This results in high airspeeds and the pilot floats down the runway as they attempt to bleed off the excess speed they are carrying as a result of the diving approach.

If you are below the correct glideslope you should increase your power setting. This will flatten the glideslope and let you achieve the correct angle. When the correct angle is achieved, decrease the power to again maintain the correct slope. A frequent error committed by pilots in this situation is pulling back on the yoke, trying to stretch the glideslope to reach the runway. The result is often landing short of the runway, landing with a high rate of descent, running out of airspeed while too high and dropping several feet, or stalling the airplane without sufficient altitude for recovery.

Any time you make an adjustment in airspeed, it will affect your glideslope. Any time you adjust your power setting, it will affect your airspeed. You must work power and airspeed in concert to achieve consistent approaches. As you are making power-setting changes, you will also be adjusting the pitch of the aircraft to hold the proper airspeed. Assuming you are at the correct airspeed, as you increase power you will need to raise the nose of the plane to prevent your airspeed from increasing, and as you reduce the power you will lower the nose to keep your airspeed from decreasing. In the first case, this will flatten the glideslope. In the second case, this will steepen it. All airplanes respond differently to power changes, so in some cases the resulting pitch change might be large, while in another aircraft it will be relatively small.

As you become proficient in your landing techniques, there should be very little need to adjust airspeed and power. Early in their flying careers, many pilots constantly change power settings and do not hold the correct airspeed, causing them to cycle between being above and below the correct glideslope. This type of inconsistent flying technique makes it extremely difficult to pick and touch down on a given point on the runway. But as you practice holding the airspeed within the desired five-knot range, you will find that your glideslopes become smoother, requiring fewer power changes.

For many light single-engine aircraft, you should be able to close the throttle while on downwind, set your airspeed, and touch down at a point on the runway you picked while on the downwind leg, without changing the power. Some higher-performance singles seem to land better with a small amount of power held on until just prior to touchdown, but the same principle applies. After setting the correct power, you should still be able to land on the target spot on the runway. We will discuss this in much greater detail in the next section of this chapter, but it is crucial to understand how to control your airspeed, how to use your power, and how changing one might require changing the other.

All of us have seen and heard of pilots increasing power and then decreasing it, not once, but several times, during an approach. First they are too low, then too high. Eventually they make the runway, then either float two-thirds of the way down it because the airspeed was too high or drop onto the runway when they reduce the power setting because they were dragging it in on the engine. This crashing-to-the-deck method might be the approved technique for aircraft carrier approaches, but not for general-aviation planes.

If you are having difficulty holding airspeed or glideslope, have your local flight instructor fly with you. A number of factors could be affecting why it is difficult for you to do this consistently. Among them could be lack of understanding of how the controls affect your pitch and, consequently, airspeed. For those who learned the concepts of airspeed control and power settings incorrectly, it could be a matter of relearning the correct technique and the reasons behind it. For others it might require getting the airplane set up earlier in the approach to allow them to concentrate more on airspeed and glideslope, instead of other things such as gear, prop, or flap settings. You might also be using incorrect airspeeds during the approach. Review the operations manual for the aircraft you fly to be sure your approach speeds are correct. Once you get the fundamentals of airspeed control and power settings down, you can begin to improve the technique you use. The best method to correct any flaws is to practice as much and as often as you can.

PICKING THE TOUCHDOWN POINT

In chapter 2 we discussed flying the pattern and each of the legs in it. In this section we are going to cover a methodology for picking your landing spot while on downwind and gauging where to make your turn onto base. At that point, we will cover flap extension points, judging the touchdown point while on final approach, and how to determine if you are on the correct glideslope. After finishing this section, you will understand the sequence of steps you should take during your approach to produce more consistent landings, touching down on a predetermined spot on the runway.

The 45° point

The classic landing approach technique has a number of steps associated with it. There can be some variations during it, depending on the type of aircraft you fly, the weather conditions, and other factors, but in this example we will start with the aircraft on downwind at pattern altitude, approximately 1000-feet AGL. While on downwind, the pilot should go through the prelanding checklist. In this example we will assume that both it and the GUMP checklist have already been completed. The pilot will pick a point on the runway to touch down on, and as the pilot approaches abeam, he or she will reduce power to 1700 RPM. (Remember to use your aircraft's recommended power settings.)

At this point it becomes necessary to reduce the airspeed to initial approach settings. Begin gradually raising the nose, allowing you to reduce the airspeed and maintain pattern altitude. Once you have reached the target airspeed, 75 KIAS for this example, you will slightly release back pressure on the control yoke. Airspeed should remain at 75 KIAS, and the aircraft will begin a descent. You should trim the airplane to hold the correct airspeed to reduce the pressure on the yoke.

Figure 4-1 shows the aircraft at a 45° angle to the touchdown point. This is known as the *key point* and is where to begin your turn to the base leg of the pattern. By using this method for judging your turning point, you reduce your dependence on landmarks near the airport for your turning points. This also makes it easier when flying at airports you are new to because you have a way to judge your turn to base that's independent of local terrain and landmarks.

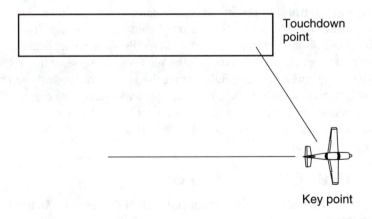

Touchdown point

Key point

Fig. 4-1. *Key point.*

After turning onto base, you can extend your first notch of flaps. You should plan your turns so that your bank angles do not exceed 30°. Be prepared for the pitch change that might take place as you extend the flaps and adjust the trim accordingly to maintain 75 KIAS. Look at the touchdown point you picked while on the base leg. If it appears to be rising, you are descending too rapidly and will touch down short of that point. If it becomes lower in the window, your glideslope is too shallow and you will overshoot that point. When the touchdown point remains fixed in height relative to the window, you will touch down on that point (except for the effects of the flare, but more about that later).

While on base, your airspeed and power setting should remain constant. You should be planning when to make your turn to final based on the wind direction and your ground speed. This is where you need to rely on the wind correction experience you have gained from performing ground reference maneuvers. If you have a tailwind on base, be sure to give yourself adequate time and space to make the turn to final. As previously stated several times, you do not want to get into a steep, low turn at a slow airspeed in an attempt to avoid overshooting the runway. This could put you in a stall that you might not be able to recover from.

Once on final you can extend the second notch of flaps and, for this example, reduce your airspeed to 65 KIAS. This is done by controlling the pitch, as previously discussed. The amount of flaps you should use will vary, depending on the airplane and weather conditions and on the type of approach you are shooting. You might need to extend full flaps once you are on short final. However, we will reserve the discussion of different approach techniques for later chapters. For now, be aware that your use of flaps will vary for different approaches.

Assuming a constant airspeed and power setting, beginning at the same point after turning to final, full flaps will cause your landing point to be closest. Partial flaps will move the point further away, and no flaps will put the landing point at its maximum distance. Given the same constant airspeed and power setting scenario, full flaps will give you the steepest approach angle, partial flaps less steep, and no flaps will result in the flattest approach angle.

As you continue on final approach, monitor your landing spot to see if it is rising or descending in the windshield. As before, if it is rising, your glideslope is too steep and you will land short of that spot. If it is descending, you will overshoot that point. If it is fixed, your glideslope is correct. When you begin your flare, you will be flattening out your glideslope. This is a necessary part of the landing process, but it affects where you will actually touch down on the runway. Because of the flare, you will touch down further along the runway from the point you were using as your landing target. If you have a precise point you want to land on, pick your touchdown target in front of that point to allow for your flare carrying you down the runway. The distance you will use varies for each aircraft and type of approach you are completing. Take some time and practice to find the right combination for the airplane you fly.

Your glidepath can also vary for each approach. Compared to a pattern flown further out from the runway, flying a pattern closer in to the runway might cause you to need a steeper angle during the descent. Holding more power during the descent can keep you higher during the approach until you reach a point where you want to descend more rapidly. Once you determine the landing point, you can choose the type of angle you want to fly. Factors that can affect your decision about the correct glideslope angle include local terrain, human-made or other obstacles near the airport, traffic in the pattern, and tower controllers.

The FAA has a series of publications as part of its accident prevention program, which I highly recommend that you obtain from the nearest FAA flight standards district office.

Publication FAA-P-8740-48, "On Landings: Part I," discusses how the path to touchdown should be viewed as a series of "imaginary windows." These windows are targets for the pilot to fly through during the approach. The first of these windows should be "flown through" just after turning to the final approach leg. The last window is the runway threshold. This is a technique to help you visualize your approach path. By having a predetermined path you want to follow down to the runway, you have a series of reference points to measure whether you are above or below the correct path. This, used in conjunction with the rising or descending landing point, makes for an effective way to determine how your approach is progressing ("On Landings: Part I," p. 2).

Figure 4-2A is an example of being too high on final approach. As you can see, the runway threshold is well down on the windshield. Early in the approach the pilot should have compensated by reducing power or extending flaps to increase the rate of descent. The downwind leg could also have been extended to allow for more time to lose altitude. Figure 4-2B represents an aircraft that has the correct altitude during the approach. The runway threshold is well situated on the windshield. Constant airspeed and glideslope will permit the pilot to continue the approach to the intended point of touchdown. Finally, Fig. 4-2C illustrates an airplane that is too low on final approach. The pilot might have flown too long on the downwind leg, reduced power too early, or extended flaps too soon. In this situation the pilot should increase power to flatten out the glideslope until the correct one has been achieved.

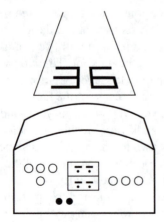

Fig. 4-2A. *High final approach.*

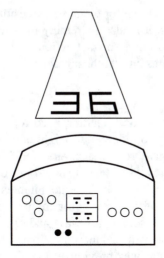

Fig. 4-2B. *Normal final approach.*

Fig. 4-2C. *Low final approach.*

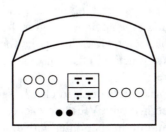

While on the topic of visualization, it is necessary to touch on visualizing the runway's position during the approach if you lose sight of it. For example, recently I was flying with a student practicing touch and goes at a controlled field. The tower controller asked us to make a tight pattern to prevent us from interfering with traffic on an intersecting runway. During the approach to the runway, the lowered wing of our high-wing airplane effectively blocked the view of the runway until we were almost onto final. In order to perform this maneuver tightly and accurately, it is necessary for you to visualize the location of the runway relative to your position. By using this technique to maintain your reference to the runway, it becomes easier to judge your turns and descent angle. If you merely start the turn and hope for the best, the results might not be what you are hoping for. In a tight pattern it is also extremely important to monitor your airspeed to prevent it from getting too slow. The distraction of flying this type of pattern makes it easy to focus on other things and let the airspeed decay.

VISION DURING THE FLARE

One of the most crucial steps in making a smooth landing is correctly judging your height during the flare. Much of the ability to judge your height is related to looking the correct distance down the runway to give you an accurate "feel" for your height above the surface. During the final approach, your vision should not be fixed on one particular point, but scanning from the aircraft's nose out to the intended landing point and back. You should also scan the far end of the runway for any traffic that might be taking off toward you, in addition to any intersecting taxiways and runways that could have aircraft that will intersect your flight path or landing rollout. The accident involving aircraft at intersecting runways that was previously mentioned might have been avoidable if the pilots had scanned a wider field of view.

Distance estimation becomes better the more you practice, but you should focus on objects that are most clearly seen. The faster your aircraft is traveling, the further ahead of the plane you should focus. As the airplane slows, you should gradually reduce the distance ahead of the plane that you are focusing on. If your focus is too close to the airplane, you will have a tendency to think you are lower than you really are, resulting in flares that are too high. If you focus too far ahead, you will be less aware of the closing distance with the ground, and it is not likely that you will flare soon enough. This often results in "wheelbarrow" landings, with tricycle-gear aircraft touching down on the nosewheel. To summarize, do not focus on just one point on the runway during your approach. Keep your eyes scanning and adjust the distance you scan as the airplane's speed is reduced during the approach. Figure 4-3 illustrates what the runway might look like if you are looking too close during the flare. This distance might vary for each pilot and plane. It takes practice and a conscious effort to establish the correct viewing habits during the flare.

There might be circumstances that make it difficult to clearly see the runway or objects around it. Since this book is concerned with VFR approaches, we will assume you are not shooting an ILS approach at minimum visibilities with clouds, fog, or other instrument conditions. However, there can be other factors affecting VFR pilots. When the sun is low on the horizon, such as at sunrise or sunset, the glare can often make it difficult to get a good view of the airport. It might be necessary to descend below the sun's horizon before you can see clearly. Some people have trouble seeing at morning or evening twilight, when runway lights become less visible. In winter regions, windows might fog up, reducing your ability to see out of the aircraft. Some general-aviation aircraft defrosters are woefully inadequate in dealing with the frost that can accumulate on the windows. On several occasions I have had to scrape a layer of frost off the inside of the front window to be able to see during the approach to landing.

Fig. 4-3. *Flare focus too close.*

Each approach can have unique circumstances associated with it. Be aware of potential visibility limiting factors as you make your approach.

HEIGHT JUDGMENT DURING FLARE

One of the most difficult skills student pilots need to master is determining the height of the aircraft during the flare. Where you focus and what you focus on are important for judging your height accurately. However, once you start to focus correctly, how do you accurately judge your height during the flare?

Dissimilar aircraft models can have substantial differences in the height that the pilot's seat is above the ground while sitting on the landing gear. These differences will affect how the correct flare height will look to you for each aircraft. If you normally fly an airplane that sits lower on its gear and go out and fly one that has longer gear, you might find yourself "driving" it onto the runway because you began your flare too low. The opposite is true if you are new to flying a plane with shorter landing gear than you are accustomed to. You might find yourself dropping the last six inches because you have established a "feel" for the correct height to flare in your usual airplane, and you need to learn the correct height for the new plane.

Several factors go into height judgment. As previously mentioned, looking the correct distance down the runway is very important in being able to judge your height. Correct use of peripheral vision is also important. While looking down the runway, I tend to monitor where the edge of the runway is in relation to the side vertical post for the front windshield. As you get down to flaring height, the runway's edge will begin to "move" up the post, giving you an indication of your relative height. It will be in roughly the same position as you touch down for each landing, so it can serve as a reference point during the flare. Runways that are wider or narrower might end up positioned in slightly different locations on the post, so be aware of the potential difference as you begin to flare.

The most important factor is practice. As you become more familiar with the "feel" the airplane has during flare, you will be able to correctly gauge your height more consistently. I find that when I fly several different models in a short period of time, I must work very hard at making consistently smooth landings in each of them. Invariably, when I switch from one model to another, the first few landings are not up to standard.

One of the common mistakes that is made during the flare is beginning the flare too late, which results in too high a rate of descent at touchdown. Another common mistake is to flare too high. In this case the plane might stall while still well above the runway, causing the plane to drop the rest of the distance to the ground and rewarding the pilot with a jarring landing. You can often tell when a pilot is not sure of the correct height to flare by how he or she "feels" for the ground. As the pilot gets down to the approximate flaring altitude, the pilot will level off, waiting to see if the airplane settles to the runway. When the airplane doesn't "grease it on," the pilot releases some back pressure on the yoke, letting the airplane settle further before leveling off again. This can happen two or three times before the pilot finally finds the correct height, and by this time the pilot might have lost so much airspeed that he or she still ends up dropping to the runway.

When everything falls into place during the flare, there will be a smooth transition as the yoke is pulled back, the rate of descent is reduced, and the airplane stalls just as the mains "chirp" onto the runway. It is an extremely satisfying moment for any pilot when the passengers comment about how smooth the landing was.

FLARING/TOUCHDOWN

We have discussed the approach, vision during the approach, and height judgment during the flare. In this section we will cover some specifics in landing conventional-gear aircraft (including wheel landings and full-stall, three-point landings) and tricycle-gear landings. We will also discuss some of the factors that can affect touchdown attitudes for tricycle-gear airplanes.

Conventional-gear aircraft

We will begin our discussion with conventional-gear airplanes, also known as tail-draggers. These airplanes have the main landing gear forward of the center of gravity and a tailwheel at the rear of the aircraft. In the early days of aviation, this was the common configuration for aircraft. It offered the ability of landing on rough, often unimproved grass runways or fields and reduced the danger of nosing the airplane over during the landing and rollout.

Over the years taildraggers acquired the reputation of being temperamental to land, with pilots "groundlooping" during landings. Aircraft manufacturers, in an attempt to make planes safer to land, developed the tricycle-gear configuration, believing they would be less susceptible to groundloops and other problems associated with conventional-gear aircraft. As a result, during the last 30 years the typical student has not had the opportunity to learn to fly in a taildragger and must look for FBOs and flight schools that offer these aircraft for instruction.

This section is not intended as a taildragger self-taught landing course, but as an introduction or review of some of the basic techniques in landing these aircraft. Before ever attempting to take off or land in a taildragger, get instruction from a qualified flight instructor. If you have not logged pilot-in-command time in conventional-gear aircraft prior to April 15, 1991, you must have an endorsement in your logbook from a qualified instructor stating your competency to fly tailwheel airplanes.

Taildragger pilots have had an ongoing argument for years about whether it is better to land in a three-point, full-stall attitude or use wheel landings, which are flown at a higher airspeed and require the pilot to land on the mains, then lower the tailwheel to the ground. The full-stall advocates state that any excess airspeed at landing is just that much more speed to damage the aircraft if you should have a problem. By landing at minimum airspeed, you are moving slower and, therefore, are safer. Full-stall landings also reduce the tendency for the airplane to nose over when landing with the aircraft in a three-point, full-stall position on soft or rough runways.

The pro-wheel-landing camp feels that full-stall landings leave you open for situations such as ballooning back up into the air due to a gust of wind, or a bounced landing, with no airspeed to maintain control over the aircraft. Wheel landings, they will tell you, let you remain in control of the plane by keeping enough speed for flying surfaces to remain effective. If trouble develops, you can still fly the airplane, not have it drop like a brick.

I don't intend to settle this dispute. Different situations and aircraft require different techniques, and you should be the judge of what will work best for you. When I first learned to fly a Citabria years ago, the instructor taught me to execute full-stall landings. This particular aircraft did not have the spring-steel gear common on newer versions of that plane today, and full-stall landings worked very well with it. I think my original taildragger instructor must have been in the full-stall corner because wheel landings were never discussed. The Citabria I currently fly is equipped with spring-steel gear, and it seems to land much better with wheel landings. The few full-stall, three-point landings I have done in it have had a strong tendency to "crow-hop" all over a hard surface runway. Full-stall landings seem to work much better on grass runways, with the softer ground reducing some of the tendency to bounce.

Right now I can hear readers saying, "If you knew how to full-stall land correctly, you wouldn't have that problem!" This could be very true. But I am using what works best for me and this aircraft. In windy or gusty conditions, I also feel that the wheel landing gives me a great deal more control to compensate for any curves the wind might throw at me during flare or touchdown. Slow airspeed, little or no rudder effectiveness, and a sudden crosswind gust that causes the airplane to weathervane could be more than any pilot can deal with. When flying a taildragger, use what is right for your experience, the type of aircraft you are flying, the runway surface, and the weather conditions. You should be capable of performing both types of landing techniques and using the correct method. If you are uncomfortable with a particular style, find a qualified flight instructor to work with you. Now that we have discussed the pros and cons of wheel and full-stall landings, let's cover some of the specifics.

Wheel landings. Wheel landings in taildragger aircraft result in the aircraft flaring as it reaches the ground and touching down on the main landing gear while the tailwheel is still off the ground. Figure 4-4 is a picture of a wheel landing as the aircraft touches down. As you can see, the aircraft is in an approximately level flight attitude.

Fig. 4-4. *Wheel landing (conventional gear).*

With a wheel landing, the approach is no different than any other aircraft. However, the flare technique is slightly different. As you begin to arrive at flare height, you need to reduce the aircraft's rate of descent. This is done by easing back on the control stick, just as with other aircraft. But since you are not attempting to achieve a full-stall landing, you do not increase the angle of attack to the same degree. The intent is to attain a zero rate of descent with the plane in a slightly tail-low attitude. As the mains touch down, the stick is eased slightly forward to firmly plant the main landing gear and let the airplane roll out, bleeding off speed. As the airspeed decays to an acceptable speed, you will gently lower the tailwheel to the runway with slight back pressure on the control stick. Be careful not to nose the plane too far forward or you might cause the propeller to strike the runway.

One thing that I like about wheel landings is the good visibility they allow over the nose of the plane during flare, touchdown, and rollout. When taildraggers have the tailwheel on the ground, the nose has a tendency to block the view ahead of the plane. In order to see where they are taxiing, taildragger pilots s-turn back and forth across the taxiway, looking out the side windows to see ahead. Wheel landings put the aircraft in a more level attitude, and the nose does not block the runway ahead until the tail begins to drop as the speed is reduced. At that point the pilot can begin s-turns as necessary to see ahead.

As long as enough speed is present for rudder effectiveness, you can use it for directional stability. Once it becomes ineffective, you will need to use brakes to control steering. Be aware that as the tail is lowered, gyroscopic precession, discussed in more depth in chapter 5, might cause the plane to have a tendency to turn right, so be prepared with either left rudder or brakes, depending on the rudder's effectiveness.

It is not uncommon for pilots new to taildraggers to be unaccustomed to the slight forward pressure used at touchdown. There is often a concern about nosing the plane over too far, causing prop damage or causing the aircraft to flip. After a few landings, this hesitancy is forgotten as the new taildragger pilot becomes accustomed to the landing procedure. The pilot must also acclimate to the sensation of being higher off the runway when wheel landing. This will require the pilot to flare at what seems to be a higher altitude than normal, even though the mains are just off the runway. In the Citabria I fly, I normally come in power off on final and don't carry any to the surface. Depending on the airplane and type of approach, you might want to carry power all the way to the ground. Whatever the approach, be sure to follow the manufacturer's recommended landing procedure.

There might also be a tendency for the tail of the plane to "wag" back and forth after touchdown. This is normally pilot induced (I know because I have done it on numerous occasions). Tail wag can be reduced by not overcontrolling the rudder or brakes after touchdown and during the rollout. Be sure not to lower the tailwheel to the runway while still traveling at a high rate of speed, or to lower it too quickly. A nice, gentle tailwheel touchdown is easiest on the tailwheel structure, and you should avoid letting it drop to the ground. After the tailwheel is on the ground, keep the control stick full back to reduce any tendency for the airplane to nose over.

Full-stall landings. In the "textbook" full-stall landing, the airplane touches down on the main gear and tailwheel at the same time. The airplane should be fully stalled as it touches down, with the control stick in the full aft position. Figure 4-5 shows a full-stall, three-point landing as the aircraft touches down. As you can see, the main gear and tailwheel will contact the runway simultaneously, and the pilot is closer to the ground than in the wheel landing.

The final approach is flown in the same manner as the wheel landing, but as you approach the runway on short final, you will need to reduce the airspeed below that

Fig. 4-5. *Full-stall landing (conventional gear).*

used in the wheel version. Because you want to end up in a full stall at touchdown, any extra airspeed will cause the plane to float down the runway before it lands, or touch down with excess airspeed, going against one of the reasons previously mentioned for executing full-stall landings.

As the airplane flares, you will increase the angle of attack to a higher value than that used in the wheel landing. As you decrease the rate of descent and reduce the airspeed, you should be just off the ground and settle to it as the plane stalls. You will find that in the same aircraft you will be sitting closer to the ground as compared to a wheel landing, so you will need to adjust your flare height to avoid "dropping in" after the stall. Since you are stalling the aircraft, you will not want to be carrying power at touchdown. This would increase the angle of attack necessary for the stall and increase the potential for touching down tailwheel first. After touchdown, the stick should be moved to the full back position, helping to prevent the tail from rising and reducing the likelihood of nosing over the plane. It is probable that the rudder will not be effective and you will need to maintain direction control with brakes, or a steerable tailwheel if it is available. As a result of the tail-low landing configuration, you will probably have very little visibility over the nose of the plane and will need to look out to the side of the plane, or s-turn, to be able to see ahead.

A common error with full-stall landings is flaring too high above the runway and dropping the plane to the ground. This can be corrected by better judging the correct height to flare, and it becomes easier with practice. Another common error is not touching down on the main gear and tailwheel at the same time. It is not unusual to see pilots land first on the tailwheel, with the mains then slamming down. Pilots also frequently misjudge the correct three-point attitude and touch down on the mains first, followed by the tailwheel. The use of flaps will change the pitch of the aircraft while on final, affecting how much pitch change will be necessary to achieve three-point attitude. This will also affect the stall airspeed. It takes practice and judgment to fully stall the aircraft just as it touches down, and, like all landings, the more often you practice, the better you will become. If done correctly, you can achieve a very gentle landing, though. I had the opportunity to fly an Aeronca Champ, and it seemed like the airplane was almost standing still in full-stall, three-point landings with just a gentle chirp of the tires at touchdown.

Landing rollout. In both wheel and three-point landings you must continue to control the direction of the aircraft during rollout. As the plane's gear touches down, friction with the runway surface generates a point for the aircraft to pivot around. The center of gravity for tailwheel aircraft is behind this "pivot point" and will have a tendency to try to move in front of the pivot. For this reason the pilot needs to anticipate this proneness and control the direction of the aircraft with rudder and wheel brakes. Tricycle-gear aircraft, discussed in the next section, have the center of gravity in front of the main landing gear, reducing the tendency of the aircraft to swerve at touchdown or during the rollout. This does not, however, mean they are completely stable and require no directional input to continue to roll straight. It is still necessary to use braking, nosewheel steering, or rudder control to keep the aircraft tracking straight. Correct control use becomes even more important in crosswind landings, which we will review in chapter 8.

Tricycle-gear aircraft

The majority of pilots today fly tricycle-gear aircraft. Originally the design was touted as safer and less demanding on the pilot than taildraggers to take off and land, but pilots still find ways to groundloop, swerve into runway lights, and run off runways and taxiways. Obviously it is still up to the pilot to safely control the airplane, and the more competent and proficient the pilot, the safer he or she will be. Do not let your guard relax merely because you are flying a tricycle-gear plane because that is when the trouble begins.

Tricycle-gear aircraft should be landed in a nose-high position, touching down on the main landing gear, with the nosegear off the ground. Figure 4-6 shows a typical tricycle-gear airplane correctly executing a full-stall landing. Notice that the aircraft's

Fig. 4-6. *Full-stall landing (tricycle gear).*

nosewheel is well off the ground as the main gear makes contact with the runway. The aircraft is approaching a full stall just as it touches down. You do not want to "three-point" a tricycle-gear plane, and you really want to avoid touching down on the nosegear first, or "wheelbarrowing." Nosegear is not designed to absorb the force of landing impacts, and more than one pilot has learned this the hard way. The results are often blown nosegear tires, leaking nose struts, or bent and collapsed nosegear.

Some aircraft tend to be more nose heavy than others, resulting in the need to be more aware of the aircraft's attitude at touchdown. Airplanes that have larger engines or retractable landing gear will have more weight situated forward of the fire wall. The result is a tendency to require more elevator travel during the flare to raise the plane's nose to the correct landing attitude. When pilots are transitioning from simple to complex aircraft, the tendency for this nose heaviness is often a surprise and requires a conscious effort on their part to pull the control yoke back soon enough and far enough to flare correctly.

Other common errors committed while landing tricycle-gear aircraft are three-point landings, with the aircraft touching down on the nosegear and mains at the same time and touching down at too high an airspeed. In the latter situation, it might be possible to raise the nose well off the runway, or cause it to balloon back into the air, as the pilot continues to pull the control yoke back after touchdown. If the nosewheel does come up, gently ease it back to the runway. Do not force it back down too quickly, as this might damage the nosegear. We will discuss ballooning situations in more depth later.

If runway length permits, it is a good idea to keep the nosewheel of the aircraft off the runway as long as there is elevator effectiveness. This generates greater aerodynamic drag, also known as *aerobraking*, helping to slow the aircraft and reducing the need for the use of the aircraft's brakes. This can result in longer brake life and reduced operating costs. Obviously this should not be done when runway length is a concern or it is not recommended by the aircraft's manufacturer.

GROUND EFFECT

When an airplane is in close proximity to the runway, the airflow around it is modified to the extent that it changes the flight characteristics of the plane. This phenomenon is known as ground effect and can be used to the advantage of pilots if they understand what it really is and how it affects the airplane. A common misconception held by many pilots is that while in ground effect, the airplane is riding on a cushion of air, causing it to float down the runway. In this section we will look at what ground effect actually is and how it affects the aerodynamic characteristics of the plane.

We need to begin with an explanation of how an aircraft flies. Lift is generated by the airfoil of the plane's wing due to reactive forces that act on the wing and airfoil as it moves through the air. A common airfoil shape used on general-aviation aircraft has a greater curvature on top of the wing and a flatter surface on the bottom (Fig. 4-7). This curvature is referred to as *camber*.

We will not get into a complete discussion of a wing's aerodynamics in this book, but will only cover the basics necessary to allow you to better understand later discus-

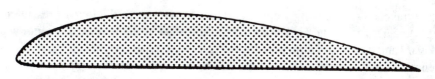

Fig. 4-7. *Airfoil.*

sions. For a more in-depth review of the topic, please read the "Flight Training Handbook," AC 61-21A. This document is published by the FAA and is available at many FBOs and flight schools.

When air meets the wing as it moves through the air, several things happen. The air traveling over the upper surface of the airfoil speeds up and, as a result, the pressure of the air over the top of the wing is reduced. The air traveling under the wing has a tendency to strike the lower surface of the airfoil, producing a higher, positive pressure under the wing. Figure 4-8 shows a typical distribution of pressures above and below the wing at several angles of attack. As you can see, the pressures above and below the wing vary a great deal, based on the wing's angle of attack.

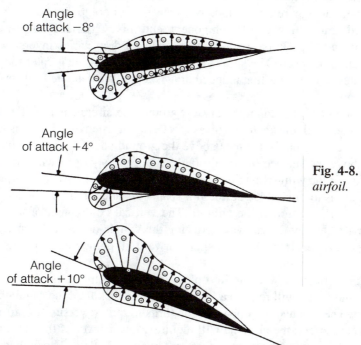

Fig. 4-8. *Air pressure on airfoil.*

There are two main forces generating lift when an airplane is flying. The first is the air that strikes the lower surface of the wing. This air is forced downward, resulting in an opposite reaction pushing the wing up. The second lifting force is generated by the air that travels over the wing's upper surface. As a result of the lower pressure on the wing's upper surface, the air leaves the trailing edge at the rear of the wing, where a backward and downward direction of flow is imparted to it. Again, an opposite reaction is generated, in this instance being forward and upward. When a wing is flying at a low, positive angle of attack, most of the wing's lifting forces are the result of the negative pressure on the wing's upper surface and the downwash at the rear of the wing.

When a wing produces lift, it also produces drag. There are two basic types of drag: parasite and induced. *Parasite drag* is the combined effects of form drag and skin friction. *Form drag* is the result of the disruption of a streamlined airflow and can be reduced by making the airplane as aerodynamically streamlined as possible during its design. *Skin-friction drag* is caused by the resistance of airflow across the surface of an airplane and can be reduced by making the skin as smooth as possible. Parasite drag increases roughly as the square of the increase in airspeed. For example, if you double your airspeed, the parasite drag of the plane will increase to four times the amount present at the original airspeed.

Induced drag is present whenever an airplane is generating lift. The wingtip vortices we previously discussed create an upward airflow near the wingtip and a downward flow behind the trailing edge. This downwash causes the vector of lift to incline aft, perpendicular to the wing's relative wind. This causes a lift component to the rear of the wing. Induced drag is this aft lift component, and this inefficiency is present when any wing generates lift. Unlike parasite drag, induced drag varies inversely as the square of the plane's airspeed. The slower an airplane is flying, the greater the induced drag it generates. This means that induced drag is highest just above stall speed, such as during landing.

As an airplane approaches to land, the airflow around the aircraft is modified by the ground's interference with the airflow. This is most evident when the aircraft is less than one-quarter of the wingspan's distance above the ground and is less pronounced at greater heights above the surface. Induced drag is reduced by 47.6% when within one-tenth the distance of the wingspan to the ground. This drops to 23.5% at one-quarter the span, and it drops to only 1.4% at one full span's distance. Figure 4-9 depicts the difference in the airflow around the aircraft. The vertical component of airflow around the wing is affected by the runway, reducing the wing's downwash, upwash, and wingtip vortices. As a result there is a smaller rearward lift component and less induced drag generated by the wing.

Several beneficial effects now come into play. To generate the same lift coefficient, a wing in ground effect will require a lower angle of attack than the same wing not in ground effect. There is also an impact to the thrust required versus aircraft velocity. While in ground effect, the airplane will require a lower level of thrust than it would require to maintain the same airspeed when not in ground effect. The higher coefficient of lift, combined with the lower thrust requirements, are what actually produce the tendency for an aircraft to float down the runway when a pilot lands carrying too much airspeed or power. The common misconception is that the cause is the "cushion of air," but it is not.

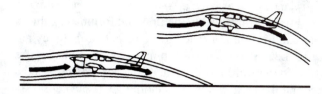

Fig. 4-9. *Ground-effect airflow.*

Ground effect also generates an increase in the local air pressure at the static source and causes a lower-than-normal indicated airspeed and altitude. During takeoff this might make it seem as though the airplane is flying at an airspeed below which it can normally fly. As the plane leaves ground effect during takeoff, several aerodynamic factors affect it. Due to an increase in induced drag and a reduction of the lift coefficient, the wing will require an increase in the angle of attack to generate the same amount of lift. The increase in induced drag will also cause a corresponding need for an increase in the thrust required. The air pressure at the static source will be lower as the plane leaves ground effect, causing an increase in the indicated airspeed. There will also be a decrease in the aircraft's stability, along with a nose-up change in moment (*Flight Training Handbook*, p. 271).

An airplane can become airborne due to the aerodynamic benefits of ground effect and might not be able to successfully climb out of it. An example would be a soft-field takeoff where the airplane leaves the ground at as slow an airspeed as possible. If the airplane is not allowed to accelerate to an adequate speed while still in ground effect, it might have little or no climb capability. Not understanding what ground effect actually is and how to use it to their advantage has caused many pilots to stall when trying to climb out of it, or to end up settling back to the runway.

UNUSUAL LANDING SITUATIONS

When everything works the way we planned during the approach, landings seem to come easily, and an uneventful outcome is taken for granted. But every now and then, Murphy's Law rears it ugly head and things begin to fall apart. Airspeeds can end up too much above or below the correct value; flaps are extended when they should not have been used; another airplane distracts you; the list is endless. As a result, we either don't notice the warning signs and correct for them, or we try to salvage the landing and end up creating an even worse situation.

In this section we are going to discuss a number of potentially bad landing situations. These include bounced landings (in both taildraggers and tricycle-gear aircraft), ballooned landings, and undershoot situations. We will begin with bounced landings.

Bounced landings

All of us occasionally bounce a landing, and normally you can let the airplane settle back to the runway with only minor recovery corrections to control it. In this section we are going to cover what actually makes the airplane climb back into the air and how to recover as safely as possible from bounced landings.

Popular belief holds that an airplane bounces back up into the air after a landing in much the same way a ball bounces after it is dropped to the floor. Unfortunately, this is not entirely true. The landing gear and tires do not rebound to the extent that they are able to force the plane to bounce to the altitude achieved in many severe bounces. As the airplane contacts the runway, the downward momentum of the airplane causes the airplane to pivot the tail down around the focal point, the landing gear, and causes an increase in the wing's angle of attack. This increase in the angle of attack results in more lift being generated by the wing, causing the airplane to climb back into the air (*Flight Training Handbook*, p. 127). The now airborne plane, lacking sufficient power and airspeed to maintain the lift, eventually settles back to the ground.

If carrying enough airspeed, the airplane can climb to a surprising altitude very quickly. If no action in these situations, the aircraft can end up dropping a number of feet to the runway.

Conventional-gear bounces. We will discuss bounced landing situations for wheel and full-stall landings in this section. Before we begin, it should be made very clear that if you are at all uncertain whether a bounced landing is salvageable, execute a go-around. We will discuss these in chapter 11, but the safer and wiser thing to do in questionable bounced landing situations is to abort the landing and try it again.

Depending on the type of taildragger you are flying, you might find it has stronger tendencies to bounce based on the type of gear it has. Some conventional-gear airplanes that I have flown had bungee shock cords that tended to dissipate the energy from a landing and not have as strong a tendency to rebound. Others I have flown had spring-steel gear that was less forgiving of incorrect landings and had a tendency to "crow-hop" along the runway when I touched down harder than I should have. These are the smaller bounces that the gear can impart, not the large bounces due to a change in the angle of attack and an increase in lift. Left uncompensated for, they can cause a plane to groundloop, drag a wingtip, or other unwanted results.

If you get into a crow-hop bounce when performing a wheel landing, ease the control stick slightly forward to help plant the mains more firmly on the runway and reduce the plane's angle of attack. If the plane bounces more than a few inches, it might be necessary to increase the throttle to help control the rate sink during the next touchdown. As you add power, you will need to compensate with rudder and ailerons to keep the airplane properly aligned with the runway. Be sure to maintain the correct landing attitude, and don't try to force the plane to stay in the air or land again until it is ready to. With wheel landings, you should touch down on the main landing gear again in a slightly tail-low attitude. As previously mentioned, if the aircraft bounces too high, go around.

In my opinion, bounces during full-stall landings present a slightly different situation. Because the airplane is at a slower speed, the controls will be less effective, and the pilot does not have as much control over the airplane until sufficient airspeed can be attained. When a conventional-gear plane bounces only a few inches during a full-stall landing, you should maintain the correct landing attitude and allow it to settle back to the runway. Depending on the attitude after the bounce, it might be necessary to pitch the plane's nose slightly up or down to achieve the correct attitude. As before, it might be necessary to increase the power slightly until touching down again to decrease the rate of descent. If you bounce too high in a full-stall landing, don't hesitate to add full power and go around. Hanging several feet above the ground in a fully stalled airplane will be an experience you would prefer to avoid, and so would the plane's landing gear.

Tricycle-gear aircraft bounces. The common practice with single-engine, tricycle-gear aircraft is to have the aircraft in a nose-high attitude at touchdown with the airspeed at or just above stall speed. From this standpoint the landing is very similar to conventional-gear, full-stall landings. Due to the relatively low airspeed, the controls will again be less effective, and you will need to stay on top of the bounce situation if it occurs.

Like a minor bounce in a conventional-gear plane, maintain the aircraft's landing attitude and let it settle back to the runway. Depending on the airplane's pitch, you might need to ease the control yoke back to slightly increase the angle of attack and lift and ease the airplane back to the runway. If the pitch is too high, gently reduce it. When you are more than a few inches above the surface, it might be necessary to add power to control the rate of sink until you touch down again. After landing, be sure to reduce the throttle to idle power to avoid using more runway during the rollout than necessary.

Under no circumstances in a bounced landing in either type of aircraft should you retract or extend flaps in an attempt to save the landing. This will cause a pitching moment in the plane and, close to the ground at low airspeed, this could quickly deteriorate into an uncontrollable situation.

Wind should also be considered when recovering from a bounced landing in both types of planes. If you have a strong crosswind, you might not be able to keep the aircraft from drifting across the runway as you descend. Be aware that what might be a recoverable bounce in low- or no-wind conditions might require a go-around with a crosswind.

Ballooned landings

If during your flare you increase the angle of attack too quickly, you will find that the airplane has a tendency to climb back into the air as a result of the increase in lift. This is known as *ballooning*. Figure 4-10 represents a ballooned landing. Like bounced landings, you can find yourself several feet above the runway in a stall. The recovery for ballooned landings is very similar to bounced landings.

If the plane balloons only a few inches, maintain the correct landing attitude and fly it back to the runway. As we have previously discussed, adjust the pitch to the correct value. Do not attempt to force the plane back to the runway too soon, as this might result in a wheelbarrowing tendency at touchdown. A common mistake is to hold the nose of the airplane too high, causing the plane to stall and drop to the runway. If necessary, add power to control the rate of descent, and if you have ballooned too high, go around. The way to avoid ballooned landings is to correctly judge your rate of descent and the timing of the flare itself. Pilots generally balloon the plane because they have

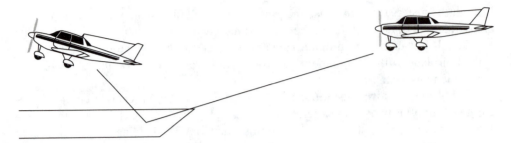

Fig. 4-10. *Aircraft ballooning.*

a higher rate of descent than they anticipated and must flare very quickly to avoid touching down too hard. As a result, they overflare and cause the plane to climb again. If you find yourself in a ballooned landing and do not feel confident trying to recover from it, execute a go-around.

Undershoot

When pilots misjudge the approach, they will occasionally touch down short of the runway, or undershoot it. This is a completely avoidable situation if a proper approach to landing techniques, as discussed in chapter 3, is used for determining your touchdown point. During the writing of this chapter, a pilot managed to land in a muddy field short of the runway at one of the airports I frequently fly from. The airplane was not out of fuel, and it was not flying in IFR conditions. The pilot misjudged the approach and undershot the runway during the landing. Fortunately, no one was injured, but there is no excuse for this in a sound airplane with no mechanical failures. If you suddenly realize you are not going to make it to the runway, execute a go-around immediately and try the approach again.

Undershoots generally result from pilots not paying proper attention to the airplane's glideslope. As a result, they have not planned and monitored where the touchdown point is. Another common cause for undershoots is when a pilot closes the throttle early in the approach. The engine cools to the point that when the pilot realizes he or she is coming up short of the runway and advances the throttle, the engine stops or experiences a loss of power. It is good practice to clear the engine by slightly advancing the throttle occasionally during the approach to keep it from having problems if you need to use it.

FLAT APPROACHES

Flat approaches are generally the result of getting too low during final approach and still trying to make it to the runway by dragging the plane in with lots of power at a low airspeed. In some cases pilots don't even add power. They just keep pulling back on the control yoke in an attempt to stretch the glide to the runway. Figure 4-11 depicts a pilot trying to incorrectly drag the airplane in rather than correctly adjusting the glideslope for a safer approach.

Flat glideslope
High angle of attack
High-power setting

Fig. 4-11. *Dragging it in.*

While you might be able to habitually fly this type of approach, there are some inherent problems with it. First, if you drag the plane in on power and the engine suffers a power loss, you can find yourself in an undershoot situation and have no way to avoid it. This situation can be compounded by flying the approach at a slow airspeed. In conjunction with a power loss, you might not be able to maintain a safe airspeed or rate of descent to the ground. If you happen to be in the camp that doesn't drag it in on power but tries to extend the glide by pulling back on the yoke, it is very possible to end up in a low-altitude stall without sufficient room for recovery.

Like the undershoot, flat-approach problems are avoidable. By properly planning your approaches, you will not need to drag the plane to the runway. If you find yourself below the correct glideslope, increase power and fly at a level altitude until the correct glideslope is intercepted. At that point reset your power and airspeed to continue the approach on the correct path to the surface. If you get too low or slow, perform a go-around and set up correctly for the next approach.

GROUNDLOOPS

Groundloops generally take place during landings and are the result of the plane pivoting very quickly around one of the landing gear during touchdown or rollout. They are usually associated with conventional-gear airplanes but are also quite possible in tricycle-gear planes. Due to centrifugal force, the groundlooping plane has a tendency to touch the outside wingtip, which can further aggravate the situation.

A number of things can precipitate the cause of a groundloop. A strong crosswind that has not been properly compensated for is a prime candidate. If the plane touches down with the lateral axis not pointed down the runway, it will want to track in the direction the landing gear is pointed. However, the momentum of the plane is down the runway, so the plane will have a tendency to pivot on the gear, setting up the groundloop scenario. Improper crosswind correction makes it possible for an airplane to weathervane into the wind and increases the chances of groundlooping. You might also have a wing lifted by a crosswind, causing the opposite wing to strike the ground and cause a groundloop. We will discuss crosswind landing techniques later in chapter 8.

Incorrect use of brakes during touchdown and rollout is another factor. I once flew a taildragger that had heel brakes. All my previous experience was in airplanes with toe brakes. The first time I landed it, I had my heels resting against the brakes too much, and when I touched down it seemed the airplane wanted to go several different directions at once. I managed to avoid groundlooping it, but it sure got my attention. The next landing, I made sure my heels were as far from the heel brakes as possible until I needed to use them.

The last cause for groundloops we will discuss is poor directional control after touchdown. Pilot-induced directional oscillation can increase with each oscillation as the pilot overcompensates trying to maintain heading down the runway. Eventually the airplane ends up groundlooping as the directional changes become too large and the aircraft loses stability. The best way to avoid this situation is to not get into the oscillations in the first place. If you do, don't overcontrol the plane, and keep your directional inputs as small as possible. If it gets too bad, go around and try again.

NIGHT LANDINGS

Night landings present a different set of challenges for pilots. Many of the visual references you become accustomed to during the day are not available during night approaches, and a different set of visual queues must be used. At night it becomes easier to spot other aircraft due to anticollision lights being more visible, but you must also rely more on your instruments and judgment when you fly. In this section we will review a number of topics associated with night landings. These will include approach planning, runway lighting, height judgment, obstacle clearance, and runway obstacles.

Approach planning

If you are one of those pilots who relies on local buildings, a grove of trees, or other landmarks for determining how to fly your pattern, you will immediately notice that at night many or all of these local "waypoints" are no longer visible. In this situation the approach methodology described earlier in this chapter becomes even more important. To illustrate night landing techniques, the following text describes the basic sequence of events that would take place during a typical night landing.

As you approach the airport, fly towards the center of it, as you would during the day. When runways become visible, you should set up your downwind entry for the appropriate runway. Distance from the runway while on downwind can be determined by using the usual reference point on the wing or strut. In this case the runway lights are aligned with it in place of the runway's edge. Approximately halfway through the downwind leg, you should turn on your landing light. This increases your visibility to other aircraft in the airport's vicinity. Abeam the target touchdown point, reduce your power, and establish the correct approach airspeed and descent.

You should continue on the downwind heading until the touchdown target, still your reference, reaches the 45° point previously discussed in this chapter. Turn to base at that point and then to final when appropriate. Use flaps as you would during a day approach and be sure to broadcast your position as necessary. The remainder of the approach will continue to be flown using standard approach settings for airspeed and power. Be sure to monitor your artificial horizon, altimeter, and airspeed to assure that your approach is correct. On overcast, dark nights, it can be difficult to use outside visual references for attitude control, so use your instruments to complement your visual scan outside the aircraft.

If the runway you are landing on has VASI available, this aids in determining if you are on the correct glideslope. If it doesn't, you should be using a touchdown point somewhat beyond the runway threshold. If that point is rising in the windshield, you will land short of it. If it is descending, you will land beyond it, and if it remains stationary you are on the correct glideslope. You might find it helpful to pick a runway light as the target reference to judge your glideslope.

Height judgment/flare

At night, as you get down to the altitude for flaring, you will notice you have a very limited field of view. Unless you happen to be flying an airplane festooned with landing lights, only a very small portion of the runway is actually illuminated. On dark nights this can offer only very limited help in determining your height above the runway. However, there are several references available to you to aid in judging your height during the flare.

The first are the runway lights along the side of the runway. In the same manner, the edge of the runway in relation to the front window side post can help determine your relative height. At night use the runway lights to do the same thing. This can be done with peripheral vision as you get down close to the runway. When the runway is illuminated by the landing light, that is a good point to start using the runway itself as a visual reference.

Some pilots also like to wait to flare until it is possible to see tire marks on the runway in the landing light. Figure 4-12 depicts how a typical runway might appear during a night approach. Depending on the type of aircraft you are flying, this might work very well in giving you enough detail to determine your height above the runway. Other pilots also use the lights at the far end of the runway. When those lights appear to rise higher than the airplane, they will begin to flare. Be aware that on long runways this might not be an accurate method for judging the flare.

Be sure to be prepared for crosswind compensation during night landings. Because of the reduced detail, it might be more difficult to determine if you are drifting in relation to the runway. Watch during final approach to determine what effect the winds are having, and compensate accordingly. We will discuss crosswind landings in chapter 8, and the same principles discussed there apply to night landings. Crab as necessary on final to achieve the correct ground track and transition to a forward slip during the flare.

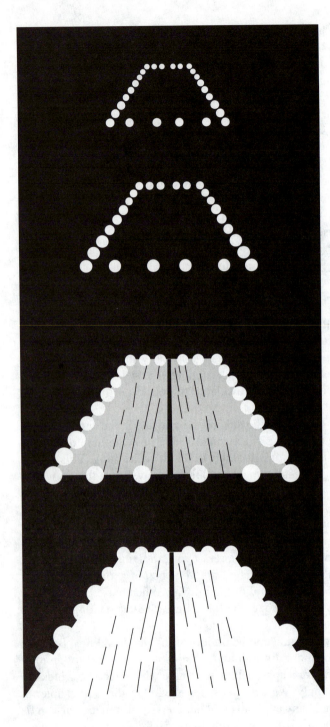

Fig. 4-12. *Night-landing runway view.*

Pilots learning night landings or those who have not executed them for some time frequently have trouble judging the correct altitude to flare and end up flaring too high or too low. The best thing you can do is to practice night landings as often as possible. If you are flying an airplane you are unfamiliar with, or have not performed night landings for an extended period of time, find a qualified flight instructor to help you become proficient. The more often you practice, the better feel you will obtain for the correct height to flare during night landings. Remember that FAR 61.57(d) states that "no person may act as pilot in command of an aircraft carrying passengers during the period beginning 1 hour after sunset and ending 1 hour before sunrise unless, within the preceding 90 days, he has made at least three takeoffs and landings to a full stop during that period in the category and class aircraft to be used." Be sure to comply with this regulation when taking passengers up to fly at night.

Runway lighting

We will briefly describe the typical runway lighting that you will find at many airports. The intent is to explain some of the information available to you from runway lights. Figure 4-13 represents lights found on most runways. The threshold lights will be

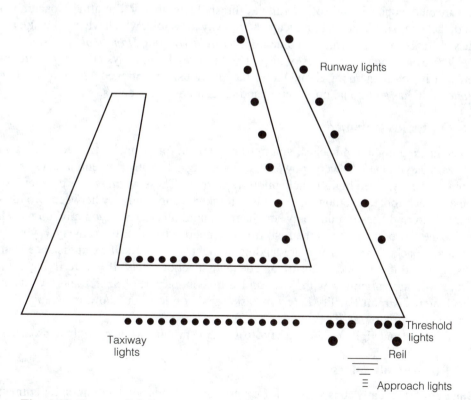

Fig. 4-13. *Runway lighting.*

green in color, and they indicate the beginning of the landing portion of the runway. There are an assortment of other types of lighting that can be placed prior to the threshold lights. Among these are runway end identifier lights (REIL), which are bright, white, flashing lights placed at the corners of the runway threshold. There are also several different versions of approach lighting systems. These are located prior to the threshold and help orient the pilot to the direction of the runway. From the air it can appear that the approach lights move toward the runway's end. This "moving" light is sometimes referred to as the "rabbit."

Lights along the side of the runway are white. There are three different classifications of these lights: low-intensity runway lights (LIRL), medium-intensity runway lights (MIRL) and high-intensity runway lights (HIRL). Some airports have lights that default to LIRL and can be stepped up in intensity by rapidly keying the airplane's mike transmit button the correct number of times. This information for an airport can be found in the Airport/Facility Directory. For instrument approach runways with HIRL, the final 2000 feet of the runway have bidirectional runway lights. As viewed from the takeoff end of the runway, this bidirectional lighting will be amber and is used to indicate to pilots that they have 2000 feet of runway remaining. Located at the end of the runway, red lights are used to show the runway end. These are also bidirectional and are on the opposite side of the runway threshold lights. For aircraft approaching to land, the other side is green to indicate the runway threshold, which we previously discussed. Taxiways are outlined in blue lights (*Flight Training Handbook*, p. 197).

There are many other types of airport and runway lighting systems. If you are interested in learning more about them, the previously mentioned "*Flight Training Handbook*" has additional information and several other references.

Obstacle clearance

When making night approaches, you must be vigilant for obstacles that could present a hazard as you land. Towers, power lines, buildings, and other potential dangers are often marked with red lights to help pilots avoid them. These warning aids do not make the approach foolproof, though. You might think you can safely fly between two towers, only to find lines running between them, or, until it is too late, you might not even notice the power lines across the road from the approach end of the runway as you fly a flat approach. Whenever you are landing at night, give yourself extra distance from obstacles and the ground to ensure an adequate level of safety. If you're flying to an airport you are unfamiliar with, you can learn about airport obstacles and layouts by checking the Airport/Facility Directory, asking other pilots you know, or asking when you call the airport on CTAF.

Planning and vigilance are the keys to avoiding problems during night approaches.

Runway obstacles

Since airports usually consist of wide expanses of land, it is not uncommon for animals to wander onto runways at night. This is even more true when the runway is warmer than surrounding ground after absorbing heat from the sun during the day. Animals will tend to warm themselves on the runway's surface. At night it might not be possible to see animals, or other obstacles, that might be on the runway. If you are unfamiliar with an airport, it might be a good idea to fly down the length of the runway at a safe altitude, with the landing light on, to verify that nothing is on the runway for you to run into during landing and rollout. After you are satisfied the runway is clear, you can climb to pattern altitude and fly a normal approach. Be sure to maintain a safe distance from the ground and any potential obstacles near your flight path.

WET RUNWAYS

If conditions are right, wet runways can present a special hazard to landing aircraft. As the aircraft touches down, a layer of water can form between the aircraft's tires and the runway surface. This layer of water prevents the tires from obtaining traction with the runway and prevents the plane's brakes from slowing the aircraft because of the very slippery surface that the plane is now riding on. This phenomenon, known as *hydroplaning*, can take place even when only a small amount of water is on the runway. Besides reducing braking ability, it can also make it difficult, if not impossible, to maintain directional control of the plane.

Today many runways are grooved to help channel water away from tires and reduce the potential for hydroplaning. However, many are not. Tire pressure is a factor in some hydroplaning situations. The minimum hydroplaning speed is the square root of the tire pressure in pounds per square inch (p.s.i.) times 9.

Landing at speeds higher than this value will result in a greater risk of dynamic hydroplaning. If you get into a hydroplaning situation, you can take several actions. One is to reduce the plane's speed to the point that hydroplaning stops. As the speed of the airplane is reduced, the tendency for hydroplaning will be lessened and the tires will begin to obtain traction with the runway, making the brakes more effective. There is no set speed at which this will take place, and it can vary depending on the weight of the aircraft, tire condition, runway condition, and a host of other factors. Once hydroplaning begins, it might exist well below the minimum speed previously mentioned ("On Landings: Part I," p. 5). Each situation is unique, and you should be prepared for it when conditions warrant. Another option is to abort the landing. If you are uncertain as to the condition of a runway, it might be best to land at an alternate airport. There is no best answer for what to do in a hydroplaning situation, but be prepared for it if you are touching down on a wet runway.

SUMMARY

In this chapter we have reviewed a number of important subjects that are all related to landing the airplane. In order to become as proficient and safe as possible, you should understand the reasons for performing certain operations in an airplane. Knowing how to consistently fly the pattern, pick your landing spot, and flare at the correct altitude can help you to smoothly and safely touch down each time you fly.

LANDING TECHNIQUES

Other topics we covered, such as bounced landings, flat approaches, and ground-loops, are all areas the average pilot hopefully does not need to recover from on a frequent basis, so it is a good idea to understand what causes them and ways to avoid getting into them. Finally, night landings should be as safe and smooth as those during the day, but you need to plan the approach accurately, know when to flare, and avoid obstacles during the approach. By doing this you can touch down on the runway as gently as during the day.

5
Normal takeoffs and landings

IN THE FOUR PRECEDING CHAPTERS, WE DEVELOPED A NUMBER OF CONCEPTS and practices necessary for safe and consistent takeoffs and landings. These were related to a variety of topics that have ranged from V-speeds to radio use, pattern entry and exit, and correct control of airspeed and glideslope. Now that we have laid that foundation, we will begin an in-depth examination of each of the different types of takeoff and landing scenarios.

As we review each takeoff and landing topic, we will cover the mechanics behind the maneuver, the rationale for actions you should take, variables such as weather that can affect your takeoff and landing, and tips or practices that I have found useful or feel should be avoided as you fly. We will also continue to discuss the differing techniques between taildraggers and tricycle-gear aircraft with each of the different takeoff and landing types. Let's begin this chapter with an explanation of factors that you should be aware of as you take off.

TAKEOFFS

Our look at normal takeoffs will touch on several areas of interest. Flap settings, trim settings, factors causing the left-turning tendency of an airplane, and conventional and tricycle-gear takeoff run and rotation are included in this section. When you are through, you should have a thorough understanding of how to correctly execute a normal takeoff, what factors can affect it, and what actions you should take to compensate for them. We will first look at flap settings.

Flap settings

Flaps are devices that are used to change the shape of an airplane's airfoil, and Fig. 5-1 depicts a number of different flap designs. Each design might be implemented for a number of reasons. Most small, general-aviation planes use the plain flap design. This is a relatively simple design to implement from a mechanical standpoint and results in a sufficient change in camber to reduce takeoff and landing speeds. In contrast, the fowler flap design is more complicated mechanically. The resulting cost and weight penalty make it less attractive for small aircraft, but it is widely used on commercial airliners. The down-and-out action of the fowler flaps can allow a greater change in the camber than can be achieved with the plain or split-flap designs. This in turn generates more lift, allows airliners to achieve acceptable takeoff and landing speeds, and permits the use of available runway lengths.

As the flap is extended, the camber of the airfoil becomes more pronounced, resulting in an increase in the amount of lift generated by the wing and the amount of drag it generates. This drag is due to induced drag and parasite drag. Remember the previous discussion of how induced drag is greatest when an airplane is in landing or takeoff configuration at slow airspeeds.

Most general-aviation single-engine aircraft have flaps that extend in increments of 10°, 20°, and 40°. However, some also have 10°, 20°, 30°, and 40° extensions. Due to the increase in lift, the airplane has a lower stalling speed and is able to fly slower than with the flaps retracted. This is of importance as the airplane is taking off or landing. For takeoffs, most manufacturers recommend no more than 10° of flaps being extended. Before using flaps on takeoff, consult your aircraft's operations manual to find the amount recommended by the manufacturer.

Having 10°, or one "notch," of flaps extended will increase the lift the wing can generate and allow the plane to become airborne at a slower airspeed and with a shorter takeoff roll. If you should happen to exceed the recommended flap setting in the belief that "if a little is good, more is better," you might find it takes more runway to become airborne than with flaps retracted or at the recommended setting. The additional flap setting will generate more drag, which reduces the airplane's acceleration rate and increases the amount of runway used to achieve takeoff speed. Due to lower airspeeds, you also might not be able to achieve as great an angle or rate of climb, preventing you from clearing obstacles near the runway's end. In fact, my first flight instructor told me

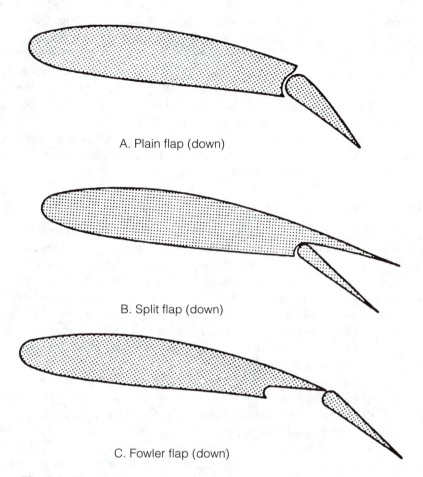

A. Plain flap (down)

B. Split flap (down)

C. Fowler flap (down)

Fig. 5-1. *Flap designs.*

that the greatest flap setting is often used to increase drag as much as to increase lift. As we will discuss later, this is useful in short-field landings, but would be counter-productive in a takeoff situation.

When flaps are used for takeoff, you will need to retract them once you have achieved a safe altitude. Be sure not to retract the flaps at too low an altitude. Most single-engine aircraft I have flown have shown a tendency to "settle" somewhat as the flaps retract. To maintain the same coefficient of lift as the flaps retract, you will need to increase the angle of attack. Depending on the characteristics of the plane you are flying, you might notice that it also has a pitching moment about the lateral axis when you retract the flaps. This is in part due to a shift in the center of pressure along the wing, and you will need to compensate for it as the flaps move from the extended position to the streamlined one. Also be aware that most aircraft are placarded for maximum speeds above which the flaps should not be out. If you exceed that speed with the flaps extended, you might cause structural damage to the flaps, flap tracks, or control mechanism. In severe cases the flaps might be completely torn from the aircraft.

I generally like to wait until I am at least 200 feet above the ground before retracting flaps. This helps to assure that I will avoid settling to the ground accidentally as I retract them. This also provides an altitude safety buffer if you are at too slow an airspeed and retracting the flaps puts you into a stall. Be aware of your airspeed as you climb and before you retract the flaps. Give yourself an airspeed safety buffer prior to retracting flaps and, as always, follow the manufacturer's recommended flap settings and procedures during a normal takeoff.

Trim setting

Elevator trim is used to reduce the control pressure you will need to maintain a given pitch attitude. For example, if you are in an extended climb, it will require less physical exertion on your part if you trim the airplane's nose up to maintain the desired climb airspeed. After trimming the airplane for a given airspeed, it will require only slight elevator inputs to adjust any minor pitch moments. Without trim you will need to supply the entire nose-up input by pulling back on the control stick. After a period of time, this can become fatiguing and an annoyance, so reduce your work load by properly trimming an airplane for the desired airspeed instead of fighting it as you climb, descend, or maintain level flight.

During normal takeoffs you should use the recommended trim setting for your aircraft. I have found that neutral to slightly nose up works well for me in most of the aircraft I fly. This requires me to raise the nose of the aircraft from the ground with a definite elevator control input. Some pilots like to have a fair amount of nose-up trim, and the airplane rotates with very little control input. I like to avoid this situation because if pilots become distracted during the takeoff roll, or have too much nose-up trim, they could find themselves in a nose-high, low-airspeed, low-altitude situation with the airplane trying to continue the climb as a result of the trim. Once clear of the ground, I trim the aircraft for the correct climb airspeed.

Conversely, other pilots like to have slightly nose-down trim during the takeoff roll. While this prevents an accidental rotation, it presents other considerations. Depending on how much nose-down trim is used, it might place additional stress on the nosegear of tricycle-gear aircraft and cause unnecessary tire wear. During experiments I have done during the takeoff run, I have noticed a strong tendency for the nosewheel on some aircraft to shimmy if too much nose-down pressure is applied. In conventional-gear aircraft it might rotate the tail off the ground before the pilot is ready and result in the propeller striking the runway. Once clear of the ground, there might be a tendency for the plane to want to pitch down, a situation to be avoided, or the plane might nose over back to the ground or settle back to the runway. There might be some exception to avoiding nose-down trim during the takeoff roll, though. In the Pitts S2B I fly, there is a noticeable difference between having one or two people in the plane and the amount of forward stick required to bring the tailwheel up during the takeoff roll. When I have a student in the front seat, I set the trim slightly more nose down to reduce the forward stick pressure required to get up on the mains. Others planes might be less affected in their trim differences as a result of differing loads. This will vary from one aircraft to another, and if you are new to a plane you might need to experiment with it to find the settings most comfortable to you.

Be sure to check and set the trim during your preflight runup. Every year pilots get into trim-induced stalls because they forget to reset the trim used during the previous landing to the correct amount required for takeoff. By using the checklist and manufacturer recommendations, this can easily be avoided.

Climb rudder

When I was learning to fly, my primary instructor drilled the use of rudder during climbs into my young, impressionable mind. He made it seem as though it was a criminal offense to let the ball move to the right during a climb. As a flight instructor, I have tried to impress on my students how important it is to use rudder to counteract the left-turning tendency of the plane. When I started learning aerobatics, my instructor commented to me how many pilots he has taught seem to just let their feet sit on the floor and use no rudder during climbs or turns. In this section we are going to discuss four factors that affect an airplane during climb and result in the left-turning tendency that airplanes with American-built engines have. This will include P-factor, torque, slip stream, and gyroscopic precession. For those of you fortunate enough to be flying a counterclockwise turning Sukhoi hot rod, or other such aircraft, this will become a right-turning tendency. The principles remain the same, however.

P-factor

P-factor is the first topic we will review. A propeller is an airfoil that is driven by the engine and rotates through the air. This generates lift, or thrust, which, in turn, results in the airplane moving through the air. As we have previously discussed, as an airfoil's angle of attack is increased, it increases the coefficient of lift. The same is true for the blades of a propeller. When an airplane is in straight and level flight, both blades of the propeller have the same angle of attack, and the same amount of thrust is created on both sides of it. As the nose of the plane is raised to a climb attitude, this is no longer true.

Figure 5-2 contains a representation of the relative difference in the angle of attack between the blades of a propeller during a climb. As you can see, the ascending blade has a smaller angle of attack than the descending blade. This results in the descending blade generating more lift, or thrust, than the ascending blade. On airplanes with engines turning in a clockwise direction, the blade on the right side of the plane will then generate more thrust, causing asymmetrical thrust and a tendency for the plane to yaw to the left. Uncorrected for, this will cause the plane to turn left during the climb, and

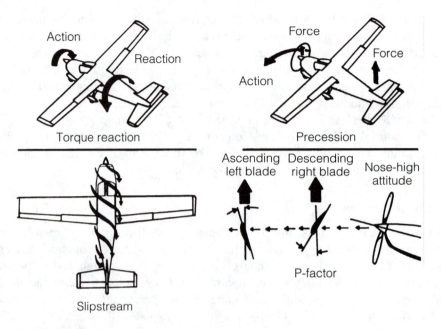

Fig. 5-2. *Left-turning factors.*

it becomes necessary to use a slight amount of right rudder to overcome the turning tendency. Remember that whenever you make rudder inputs, you will need to make a corresponding aileron input to maintain a wing's level attitude.

In a descent, the opposite is true. The blade on the left side of the plane now is at a greater angle of attack than the right, and it now generates more thrust. This will result in the need for a slight amount of left rudder to prevent the unwanted yaw to the right from taking place. You might find that in a descent you are at lower power settings and airspeeds, so the yawing tendency might not be as pronounced, but you should be aware of it and compensate for it as necessary. When doing aerobatics, I have noticed in some descending portions of maneuvers that it is very apparent that the left rudder is required to hold a straight line. If you are uncertain of the use of controls during a climb or descent, find a competent flight instructor and have him or her demonstrate correct methods of correction for P-factor.

Torque

Sir Isaac Newton stated that, "For every action there is an equal and opposite reaction." This law of physics sets the stage for our next topic of discussion, torque. As an aircraft engine turns the propeller, it also imparts a rolling motion in the opposite direction to the aircraft itself. Because the prop is smaller and lighter, it does the majority of the turning, but the rolling moment that is induced in the airplane causes it to tend to roll to the left.

This rolling tendency will be more pronounced at high-power settings at low airspeeds, when the engine has highest torque and there is less airflow over the wings and control surfaces to overcome the rolling tendency.

Figure 5-2 also depicts the action/reaction concept as related to torque. It is evident how the airplane is trying to roll in a counterclockwise direction as a result of the propeller turning in a clockwise direction. As with P-factor, right rudder is necessary to offset the effects of torque to keep the airplane tracking in the desired direction (*Flight Training Handbook*, p. 33).

Slipstream

As the propeller turns, it generates a "corkscrewing" body of air, or slipstream, that turns in the same direction as the prop, in a clockwise motion. Figure 5-2 illustrates how this air flows around the fuselage of the plane and strikes the vertical stabilizer/rudder on the left side. The air is, in effect, pushing the tail of the plane to the right, and this causes the nose to move to the left. This is the third left-turning tendency and it, like the previous factors, is counteracted with the use of right rudder. Like torque, this tendency is strongest at high-power settings and slow airspeeds. At slow airspeeds, the slipstream is shorter and has more force as it strikes the vertical stabilizer. As the airspeed increases, the slipstream lengthens and lessens in the strength with which it strikes the tail (*Flight Training Handbook*, p. 33).

Gyroscopic precession

The last left-turning tendency we will discuss is gyroscopic precession. To begin, we will review what a gyroscope is and some characteristics it has, then we will relate this information to the propeller on a plane. You might recall from your high-school physics class that a gyroscope consists of a rapidly spinning disk or wheel balanced on an axle. Due to the properties of the spinning disk, it resisted movement from its original orientation. This resistance to movement make it ideal for use in instruments requiring a platform that maintains its original position relative to aircraft movements. As a result, the artificial horizon, direction gyro, and turn-and-bank indicator each contain a gyroscope.

The propeller also acts like a gyroscope. It becomes a rapidly spinning disk and possesses the same properties of other gyroscopes. It resists movement like all gyroscopes and also possesses the property of precession. Precession is the physical reaction a gyroscope has related to its movement. Briefly stated, precession causes the result of a force being applied to a gyroscope to take place 90° in the direction of rotation from the point at which it was applied.

Figure 5-2 depicts a clockwise spinning disk. In this example a force is applied at a point at the top of the disk, attempting to push the top of the disk forward. As a result of the previously stated physical reaction, the action actually takes place at a point 90° from the point of the force application, in the direction of rotation. As a result, the disk pivots to the left, as opposed to pitching the top forward.

Gyroscopic precession is most prevalent at high-power settings, such as during takeoff. When you raise the tail of a conventional-gear plane during the takeoff roll, prior to liftoff, it is the same as applying the input force at point A in the example. As a result, the aircraft has a tendency to yaw left and requires right rudder to compensate. If you pitch the nose up, gyroscopic precession will have a tendency to cause the nose to yaw to the right. The more "rapid" the pitching or yawing moment, the stronger the effects of gyroscopic precession (*Flight Training Handbook*, p. 33).

Manufacturer's compromise

Undoubtedly you have asked yourself, "If these forces are always present, why does the plane I fly maintain straight and level flight once I level off at altitude without needing to compensate with rudder continuously?" The aircraft manufacturer has designed the aircraft to offset the effects of the four factors during straight and level flight. In order to rig the plane to fly straight and level, several design concepts were incorporated into the airframe and engine. As a result, after you level off at straight and level cruise flight, these design features offset the four left-turning tendencies the plane has. We will discuss several of the design features to make you more aware of how the airplane is rigged to help reduce your work load as you fly straight and level. You might not be able to physically see the changes because in many cases they are very subtle, but the results become evident as you fly. We will cover engine offset, vertical stabilizer offset, and wing washout.

On most single-engine aircraft, the engine is canted slightly to the right. This produces a thrust line that is slightly offset from the longitudinal axis of the plane and causes a small right-turning tendency. Engine offset is normally accomplished by canting the engine mount and might require only a few degrees of offset. Lower-horsepower engines are usually offset less than higher-powered engines because more powerful power plants generate a stronger left-turning tendency, and more offset is required to compensate. Higher-power settings also produce a stronger left-turning tendency than lower ones. The manufacturer has designed the offset to work best at cruise power settings, where the airplane will spend most of its time.

In much the same way the engine has been offset, so has the vertical stabilizer. By angling it to the longitudinal axis, the airplane uses the vertical stabilizer to produce yet another right-turning tendency to offset the left-turning factors. Like the engine, the vertical stabilizer's offset will vary from one design to another. Higher-performance planes might require more offset than lower-performance ones, and this offset produces the desired amount of right turn at cruise. It is less effective in other flight regimes and is used in conjunction with the engine's offset to help the airplane fly straight during level cruise flight.

The last design factor we will cover is wing washout. *Washout* is the slight twisting of the chord line of the wing along its length from inboard to outboard and is used by the manufacturer to give the plane more desirable characteristics in different flight configurations. Normally the chord line is twisted to produce a slightly lower angle of attack at the wingtips than at the wing root. As a result, the wing will stall first at the root and still produce lift further outboard, which allows the ailerons to remain effective in a stall. To help overcome the left-turning tendencies, each wing might contain slightly different washout so that the left wing generates more lift than the right wing. This causes a tendency to roll to the right and helps to overcome the left-turning tendency. Like engine and vertical stabilizer offset, the differential washout between the left and right wing is designed to work optimally when the airplane is in cruise flight.

The aircraft designer blends engine offset, vertical stabilizer offset, and wing washout together to produce a desired set of flight characteristics. If only one method were used, it could result in adverse side effects that make the airplane more difficult to fly. Next time you preflight your airplane, see if you can notice the slight offset angles or differences in washout. You might not be able to see them, but understanding these left-turning tendencies and offsetting design features can help you maintain better control as you fly the plane.

Different airplanes also react differently and require differing amounts of right and left rudder to counteract these forces during climbs and descents. Depending on the horsepower of the engine, size and weight of the propeller, how the aircraft is rigged, airspeed, and a host of other factors, you might find that different amounts of right rudder are required in different aircraft. The same is true of left rudder in descents. The Cessna 182 I fly for skydivers requires almost no right rudder to keep the ball centered during climbs, but it needs a generous amount of left rudder when descending on final approach. I suspect that since it spends most of its time in a climb mode, other pilots have adjusted the ground-adjustable rudder trim to compensate for the right rudder needed for climbs. The Citabria I fly is rigged very well and requires right rudder during climbs and left rudder in descents, behaving the way a correctly rigged airplane should in different flight modes. By comparison, the Pitts S2B has a very sensitive rudder and lots of horsepower. In a climb, it needs right rudder to compensate for left-turning tendencies, but the real surprise is when you pull the power back. The left rudder really comes into play as power is reduced to keep the plane from yawing to the right. You should build a feel for each plane you fly and understand how much you need to compensate with rudder to maintain coordinated climbs and descents. Remember that any time you make any rudder inputs, you will also need to make a corresponding aileron input as well.

Conventional-gear ground roll/rotation

Now that we have discussed the factors that affect the plane as you take off and climb, let's begin to look at what is involved in the ground roll and rotation of a conventional-gear airplane under normal conditions. Later chapters will cover techniques for crosswind, short- and soft-field takeoffs (see chapters 6 and 8).

However, in this section assume that there is no crosswind and you are taking off from a hard-surface runway of sufficient length. We will also assume that you have completed the pretakeoff checklist and obtained the necessary clearance for takeoff, if necessary.

Before you roll onto the runway for takeoff, glance around the pattern and verify that there is no other runway traffic that could pose a potential for compromised safety. There have been several occasions during my flying career that, while on final, another plane taxied onto the runway while I was on short final. On two instances I had to execute a go-around and alter my flight path to assure that I did not land on top of the aircraft taking off or collide with it after it lifted off. I altered my course to the right so that I could maintain visual contact with the aircraft after it left the ground. Be sure to look above and below the normal glide angle you think planes will use. Some aircraft, like the Pitts S2B, could be at 1000 feet on short final due to the steep glideslope they have when power is pulled for landing. Also look at intersecting runways to be sure that no other aircraft that might pose a problem are taking off or landing. Never assume that the runway or pattern is clear just because you do not hear anyone reporting over CTAF.

As you roll onto the runway, turn to align yourself with the runway centerline and do a last-minute check of trim, flap, and engine instruments. Some tailwheel aircraft are more "blind" over the nose than others and remain that way until the tail begins to come up. If you start the takeoff roll not pointed down the runway and you don't find out until the tailwheel rises, it might be too late to recover. Ailerons, elevator, and rudder should be in neutral positions. Once you are satisfied that everything is in working order, advance the throttle smoothly to full power. You will very likely need to apply a certain amount of right rudder to overcome the left-turning tendencies we previously reviewed. More powerful engines might require more right rudder, but this will depend on several factors, and you should be prepared to use the correct amount of rudder. During power application, the control stick should be in a neutral position and as the plane accelerates, you should add a small amount of forward stick. Once sufficient speed has been attained, the plane's tail will come off the runway. As the tail comes up, precession might cause an additional left-turning moment and require more right rudder in order to keep the plane tracking straight down the runway.

The plane should be in approximately level flight attitude as it continues to accelerate. In the Citabria I fly, at approximately 60 mph I apply back stick and rotate the plane from the runway, transitioning to an 80-mph climb speed. In most airplanes, very little back pressure on the stick is required to rotate. The plane should achieve the proper climb attitude to produce the correct airspeed as you leave the runway and not require you to apply varying pitch inputs as you "hunt" for the right airspeed. In normal takeoffs, I like to use the recommended takeoff climb speed or the best-rate-of-climb speed. The recommended climb speed might be higher than the best-rate-of-climb speed, resulting in a lower angle of attack. This helps promote better engine cooling during high-power-setting operations. If your elevator trim is set correctly, the plane should require very little back pressure to maintain the climb airspeed. Be careful not to apply too much elevator back pressure, which could cause the plane to overrotate. If this happens, the airplane might lose airspeed and stall or settle back to the runway. Should you find yourself in this situation, gently ease the nose down to allow the plane to pick up airspeed.

Avoid overcompensating and do not lower the nose too much or too quickly, as this could result in the plane nosing over onto the runway or settling back down to it. Either situation could damage the airplane, or worse. Use firm, appropriate control inputs when you fly. You are in control of the airplane and should make it fly within its and your capabilities. Overcontrolling, undercontrolling, or indecisiveness can cause or aggravate a bad takeoff or landing situation. Know your airplane's flight envelope and characteristics and be prepared for the unexpected.

One note concerning the rotation method we just discussed. If your airplane's manufacturer recommends a different technique, use it. This method has worked for me in several different makes and models of conventional-gear planes, but there might be those for which it will not work. Some warbirds have props on them that are so large in diameter it would strike the runway if the plane was brought up to a level flight attitude. These planes require lifting off in a three-point attitude. This is an extreme example, since very few of us have the opportunity to fly these types of planes, but there might be other factors that require lifting off in a three-point attitude for a normal takeoff.

As you leave the runway, you will probably need to input more right rudder to offset the increased right-turning tendency present in a climb attitude. As you climb out, monitor the engine and flight instruments, but also remember to scan for traffic in the airport's vicinity. If you happen to be flying a retractable-gear aircraft, use the manufacturer's recommended procedure for gear retraction. However, a general rule of thumb is that when you reach an altitude and distance down the runway where you would not be able to land on it again, retract the landing gear. Be sure not to exceed the plane's gear-down or gear-cycling airspeeds while the landing gear is down. At this point you can use the procedures we have previously discussed to exit the pattern.

Frequent mistakes made by pilots during takeoff in conventional-gear aircraft include not compensating with the correct amount of right rudder prior to raising the tail from the ground, not raising the tail to achieve the correct attitude during the ground roll or at the correct airspeed, and not tracking straight before, during, or after the tail is raised. Since most tailwheels are castering, nonsteering, or only marginally effective for steering, you might need to rely on the use of brakes to maintain your track at the start of the takeoff roll, and rudder once sufficient speed has been reached. If the plane you fly has a steerable tailwheel, this might prove to be sufficient, and little or no brakes might be necessary for directional control during the takeoff roll. Many pilots have a tendency to overcorrect with the brakes and rudder during the takeoff run, and the plane might develop a tendency to oscillate to the left and right. Keep your control inputs small and avoid the tendency to use too much brake or rudder. Anticipate what the plane is going to do as much as possible and what control inputs you will need, then smoothly apply them. This will help avoid the overcontrolling tendency. If you should happen to begin to overcontrol, relax and let the airplane stabilize. Then begin again with smaller inputs. Once the tail is off the ground, keep in mind that now you are balancing the plane on two wheels that are ahead of the center of gravity. If you use brakes too heavily to steer the plane, it might have a tendency to nose over, so again, do not overcompensate when you apply brakes.

If you do not raise the tail to the proper takeoff attitude, you might find the airplane takes longer to achieve the correct rotation speed or that it becomes airborne without your rotating it. The reason it might take longer to accelerate is there will be more drag with the wing or wings at a higher angle of attack. This drag will reduce the acceleration of the plane and increase the ground roll. By attaining an approximately level flight attitude once you raise the tail, you will reduce the angle of attack and drag. Not raising the tail high enough might cause you to become airborne without a rotation, again due to the increased angle of attack. You will remember from previous discussions that increasing the angle of attack increases the lift generated by the wing. Maintaining a tail-low attitude during the ground roll could generate sufficient lift to cause the plane to take off before a safe airspeed has been reached. Due to ground effect, the plane might leave the ground, only to settle back down or stall once enough altitude has been gained to reduce ground effect. If you should happen to take off before you have adequate airspeed, reduce the back pressure slightly and stay in ground effect until your speed increases to the correct value. Avoid settling back to the runway or stalling the plane.

Tricycle-gear ground roll/rotation

The ground roll and rotation for a tricycle-gear airplane is very similar to that used by conventional-gear aircraft. After performing the recommended runup and clearing the pattern for any other aircraft, you will turn onto the runway centerline. Once aligned with the runway, advance the power. You should begin with the ailerons, elevator, and rudder in the neutral position. As full power is applied, I like to quickly scan the engine gauges to assure that the oil pressure and other gauges are within the correct ranges. As you apply power, you will likely need to apply sufficient right rudder to correct for the left-turning tendencies we have previously discussed. Unless you happen to be flying a tricycle-gear plane that has a castering or nonsteerable nosewheel, you will use the rudder pedals, which are attached to the nosewheel, to maintain directional control. Unlike taildraggers, you will not raise the tail. You will remain in an approximately level attitude as the plane accelerates.

At the airspeed recommended by the manufacturer, you should ease back on the control yoke and raise the nosegear from the ground. The attitude should be approximately that used for climb attitude. This helps to achieve a smooth transition from takeoff to climb. As you raise the nose, the greater angle of attack might require some additional right rudder to maintain a track parallel to the runway. Depending on the plane, horsepower, rotation airspeed, and other factors, the airplane might roll along the runway on the mains with the nosewheel off the runway. At this point you will need to maintain directional control with rudder and/or brakes, since nosewheel steering will no longer be effective. As the airspeed increases, the plane will lift off from the runway and continue to climb.

As we discussed with conventional-gear planes, use control inputs that are as small as necessary. If you find yourself in a situation where you are overcontrolling the aircraft, let the plane stabilize and begin again with smaller control inputs. The sooner you realize that an overcontrol situation is beginning and you correct your control inputs, the easier it will be to recover. If the situation gets too far out of control, it might not be possible to stay on the runway or recover gracefully from the situation. If you do run off the sides or end of the runway, do your best to avoid runway lights and other objects that could damage the aircraft.

Once the situation has deteriorated to this point, there is no textbook answer for correct recovery. However, you will probably want to pull the power back and get the plane stopped as soon as possible. I have talked to and read about pilots who managed to get the plane airborne and fly out of the situation, but this is not something I recommend. Because of the uncontrolled state the plane is in, you might end up dragging a wingtip, collapsing a gear, or striking an object on the ground as you take off. Due to the fact you would be attempting to get the plane off the ground at what will likely be a lower-than-called-for airspeed, you might end up stalling the aircraft and impacting the ground again. If you find yourself in this type of situation, keep your head, assess the situation, and take the appropriate actions. Always maintain control of the plane and think about what you need to do to prevent damage to the plane and injury to yourself and any passengers. Prevention is the best answer, though, so get the plane correctly set up for the takeoff roll from the start, and use proper control inputs.

Some of the common errors seen during takeoff roll in tricycle-gear planes include raising the nose from the ground too quickly, not using right rudder to compensate for the left-turning tendency, raising the nose too steeply as the plane leaves the ground, and waiting too long before raising the nosewheel. In a normal takeoff, the nosewheel should be left on the runway until the airspeed recommended by the manufacturer is achieved. Raising it too soon in the takeoff roll could result in the plane leaving the ground before sufficient speed has been attained. The plane could then stall or settle back to the runway after liftoff and cause damage to the landing gear or other parts of the airframe. Not using right rudder could result in the plane veering to the left and running off the side of the runway. Raising the nose too steeply after liftoff could result in the plane stalling and settling back to the runway, again, potentially causing damage or injury. Finally, waiting too long to raise the nosewheel could result in the nosegear shimmying due to its remaining on the runway at speeds higher than it was designed for. In some cases the shimmy could become quite severe and cause loss of control or damage to the gear itself.

LANDING

In this section we will discuss landing procedures for normal landings in tricycle- and conventional-gear planes. We will assume you have correctly entered the pattern and have completed the prelanding checklist for your aircraft. We will also assume that if you fly a retractable-gear plane, the gear is down and locked. Topics of interest will include flap settings, trim settings, approach rudder, and tricycle- and conventional-gear flare, touchdown, and rollout. We will also review some of the common mistakes made during landings.

Flap settings

Flaps are used to help reduce the stall speed of the aircraft. As a result, the plane can be safely flown at slower airspeeds during the approach. The combined effects of slower airspeeds and the drag developed by the flaps allow the plane to land and stop in a shorter distance than would be possible without the use of flaps.

In a normal landing situation, where it is not necessary to maintain a steep glideslope to clear an obstacle at the end of the runway or to get the aircraft stopped in the shortest possible ground roll, the use of flaps should be in accordance with the recommendations of your aircraft's manufacturer for normal approaches to landing. In many cases this involves the use of one notch or 10° of flaps. This gives you the ability to reduce the airspeed to a more manageable one that decreases the ground roll and landing stresses on the plane, yet does not put you in the steep glideslope that is associated with short-field landings. Remember that many aircraft have a pitch change associated with the extension of flaps, and it might be necessary to adjust the pitch and trim to maintain the correct airspeeds. A common error that occurs when flaps are extended is that the pitch change is not compensated for and the airplane begins to slow too much.

Trim settings

Trim settings become important when you want to fly consistent airspeeds during the approach. The correct use of elevator trim can reduce the work load and level of physical effort necessary to maintain a constant airspeed. For example, if you enter the pattern and reduce power on downwind to the appropriate setting, you will then hold pattern altitude until the airspeed is reduced to the correct value. At this point you will be pulling back on the control yoke, receiving no help from elevator trim. As you continue on and turn to base, you will extend a notch of flaps, with another change in elevator control pressures to hold the desired airspeed. Finally, as you make the turn to final and reduce your airspeed still further, you will again physically compensate for the loss of elevator effectiveness.

This can become a physical and mental drain, depending on the circumstances you are flying under. If your attention becomes diverted, you might relax the back or forward pressure you are applying to the elevator and the airspeed, and glideslope will fluctuate. By correctly trimming the elevator to reduce the elevator control pressure to a more neutral feel, it requires less effort on your part to hold a given airspeed. When you extend flaps, you might need to compensate until you retrim the elevator, but in most cases this will require less effort than not having already made a trim adjustment. Properly trimmed, the plane will tend to hold the desired airspeed on its own, and momentary diversions in attention will not be as likely to result in airspeed fluctuations. Correctly trimming the plane during the approach will also reduce the amount of back pressure necessary during the flare.

When flying normal approaches with only one notch of flaps, it is likely you will require only small changes in trim settings to arrive at neutral elevator inputs. This will vary from one plane design to another but should generally hold true. Try to retrim the plane as soon as possible after the power is first reduced in the pattern and again, if necessary, after the flaps are extended. To test for the correct trim setting, you should be able to let the control yoke or stick "float" in your hand, and the proper pitch attitude and airspeed will remain constant. If you find that you need to apply forward or back pressure because the nose is rising or descending, you need to adjust the trim in the correct direction. Make small trim changes and let the plane stabilize before deciding if the new setting is correct. After a few approaches, you will develop familiarity with the plane you are flying, and it will require less experimentation to get the trim setting right.

Approach rudder

The same factors that affect your plane during takeoff also change the flight characteristics during your approach and landing. Due to the reduced power settings, they might not be as prevalent during the approach, but you might need to compensate for them to maintain coordinated flight as you land.

Torque's left-turning tendency will be less when power settings are reduced as compared to during takeoff while you are under full power. The same is true of the slipstream. With the propeller turning slower and with less power, the slipstream will not have as strong an impact on the left side of the vertical stabilizer and rudder. Because you are pushing the nose down, gyroscopic precession will be translated into a left-turning tendency. But, like the other factors, with reduced RPMs it will not create as much left yaw.

One factor that tends to create a right-turning tendency in a descent is the propeller. While descending, the downward-traveling blade of the propeller now has a lower angle of attack than the ascending blade. As we previously discussed, in American-made engines that turn clockwise as seen from the cockpit, the ascending blade will now generate more lift and cause a yawing tendency to the right. Depending on the type of plane you are flying, its propeller, and a number of other factors, when you are in a descent you might need to use left rudder to hold coordinated flight.

For example, in the Cessna 182 I fly for skydivers, I have noticed a stronger need for left rudder during a descent than in many other aircraft I fly. This might be due in part to the way the plane is rigged and the higher-horsepower engine it has. Some of the trainers I teach in require little or no left rudder during a descent, but, like the 182, I have noticed a similar need for left rudder in the Citabria I fly. The most dramatic need for left rudder I have encountered during reduced-power descents is the Pitts S2B. This plane has so much propeller mass, horsepower, and torque that when power is pulled back, there is a very strong tendency to yaw to the right. When flying tight, left patterns in this plane, a large amount of left rudder is needed during the turns to maintain coordinated flight.

The plane you fly might respond differently from any of the aircraft I have described. However, you should make a conscious effort next time you fly to understand the forces that affect the plane and how much rudder you need, if any, to compensate for the right-turning tendency that might be present during landing. If you have ever wondered why it seems the plane feels uncoordinated during approach, it might be that you are yawing and do not realize it. By using the correct amount of rudder, you will eliminate the slipping or skidding tendency that might be present.

Conventional-gear flare/landing rollout

When making conventional-gear, normal landings without crosswinds, the procedures will be very close to those we covered during discussions about landings in tricycle-gear aircraft. Depending on the type of tailwheel aircraft you fly, you might need to adjust the technique for that plane, but the basics will remain substantially the same. In this section I will include examples for the two tailwheel planes I currently fly, a Citabria 7KCAB and a Pitts S2B. While both are conventional gear, they have very different requirements during landing.

When you are abeam the landing point on downwind, reduce power to the manufacturer's recommended setting for your plane. At this time you should begin to reduce your airspeed and trim the plane to maintain airspeed. At the 45° "key point," begin your turn to base and extend the recommended flaps for normal approaches. None of the taildraggers I fly have flaps, so in my case this step is skipped. As you turn on to final, you should be lined up with the runway centerline and holding the correct airspeed. Since this landing is in no-wind conditions, or with the wind directly down the runway, you will not need to set up a slip or crab to remain on the correct ground track to the runway.

In the Citabria and Pitts, I like to make standard approaches with power off. This helps to provide a steeper rate of descent and provides two benefits. The first is that by flying power-off approaches, if the engine fails you already have planned your landing with the engine at idle and should still be able to glide to the runway. The second is that with a steeper approach, visibility over the nose is better. This is especially true of the Pitts S2B, where during the approach the entire runway can disappear if you are forced to fly a standard pattern by other traffic or a tower. Any time I need to fly an extended final with that plane, I stay high until I have the runway made and pull the power. Depending on my altitude and the length of the runway, I might need to do s-turns on final to keep the runway in sight. Once the power comes back, I can lower the nose of the plane and see the runway fairly well.

Whether you are executing a three-point landing or a wheel landing, your ailerons and rudder should remain in approximately the neutral position during the flare and touchdown, and the plane should be aligned with the runway centerline as you flare. In a wheel landing, the runway should be visible forward over the nose of the plane. Standard flare sighting techniques of scanning from the nose of the plane out to the intended landing point and back should work well in helping you to judge your height during the flare. When making three-point landings, visibility over the nose might be reduced or lost completely, depending on the plane you are flying. If this is the case, you will need to use peripheral vision during the flare to judge whether you are in the correct three-point attitude and to judge your altitude during the flare itself. With both landing techniques, your plane should touch down gently on the runway with a minimum rate of descent.

We have previously discussed landing rollout, but let's review some points as they are related to a standard landing. If you are executing a three-point landing, your ailerons should remain in a neutral position. If you do not touch down with both mains at the same time, or if you are rolling on one main and the tailwheel, with the other off the ground, it is likely you have given some aileron input during the approach. This technique is used in crosswind landings but should be avoided in noncrosswind situations. When airspeed dissipates to a safe point, you should bring the control stick to the full aft position to help reduce any tendency for the plane to nose over. Be careful not to bring the stick back too quickly or with too much airspeed, though. This could result in the plane lifting off the ground again with little or no airspeed, altitude, or ability to recover gracefully.

Once on the runway, you will very likely need to work the rudder pedals and brakes to maintain your track down the runway. In a wheel landing, the tailwheel will remain clear of the runway until you lower it to the surface. As we previously discussed, you will need to carefully use rudder and brakes to continue tracking straight. As you lower the tail to the runway, you will probably have less vision forward, so you will need to use peripheral vision as you continue down the runway. In the three-point position, you will already have the tailwheel on the ground when you land.

Different airplanes can have strikingly different handling characteristics during the rollout. The Citabria I fly tracks very nicely down the runway and requires very slight rudder inputs in no-wind landings to keep it straight. Vision over the nose remains good even after the tailwheel is on the ground, and normally only rudder is required with minimal brakes. The Pitts S-2B is much different during the rollout. I land it in a three-point position, and after touchdown it goes through what is known as the *Pitts shimmy* or *Pitts wiggle*. As the weight transfers from the wings to the landing gear, the plane wiggles and rocks. This requires small inputs on the rudder pedals to keep the plane tracking straight, and if inputs are too large it can make controlling the plane more difficult. However, once you get used to the transition, it becomes second nature to anticipate and adjust for it. The plane you fly might have its own unique landing characteristics that you must be aware of. When landing in no-wind conditions, the basic control inputs we have discussed should work in most tailwheel airplanes, but be sure to use those that work for the plane you are flying.

Tricycle-gear flare/landing rollout

In this section we are going to review the flare and landing rollout for tricycle-gear planes. Much of what we discussed in landing and rollout of conventional-gear planes holds true in this section. We will refrain from reviewing the pattern itself since that was just covered, but will instead move on to short final approach.

In no-wind conditions, the longitudinal axis of the plane should be lined up with the runway when you are on final. Your airspeed should be constant, and no crab or slip will be necessary to maintain your alignment with the runway centerline. As you begin the flare, ailerons should be neutral, and little or no rudder should be required. This last point could be affected by your plane's need for left rudder during the descent and its need for right rudder as the nose is raised during the flare. In most tricycle-gear planes I have flown, visibility over the nose is usually quite good, so your ability to see the runway during the flare should not be a factor. If the plane you fly does restrict visibility during the flare, the techniques discussed in the previous section might be useful in seeing as much of the runway as possible during touchdown.

After touchdown on the main landing gear, you should hold the nosewheel off the runway and gently lower it to the ground as your speed dissipates. This reduces stress on the nosewheel as compared to letting it drop to the ground immediately after landing. Assuming you have nosewheel steering, as the nosewheel settles you will have steering capability and be able to use it to track straight down the runway. Like tail-draggers, inputs to steering should be only as large as necessary to keep the plane tracking straight. If you find that you are touching down on one main landing gear in no-wind conditions, then you probably have unwanted aileron input during the flare. This is actually not uncommon for some pilots who tend to add a twisting motion to the control wheel as they pull back during the flare. If you find that this is the case, make a conscious effort to keep neutral aileron input as you flare. After very little practice, this will become the normal feel you develop during the landing.

Once most tricycle-gear planes are planted on all three gear, they tend to be fairly stable during the rollout. In no-wind conditions, you might find that little or no steering input is needed until you turn off the runway. Do not let yourself be lulled into a sense of complacency, though. Keep your attention on the airplane and controlling it. Even though tricycle-gear planes have developed a reputation as being more stable to land than conventional-gear planes, they still demand your attention. If you start to ignore controlling the plane and start thinking ahead to something else, you might find yourself headed for the edge of the runway. No matter what type of plane you are flying, landing, or taxiing, stay on top of it from the time you start the engine until you park it and shut it down.

SUMMARY

In this chapter we have reviewed the takeoff and landing techniques for both tricycle and conventional-gear planes in no-wind or straight-into-the-wind landings. After reading this, you should understand the control positions that will be necessary in a textbook landing situation. In reality you might find that absolutely straight-on headwinds or dead-calm days are few and far between. However, if you understand the basic control positions for normal takeoffs and landings, then this becomes the building block as you move on to other types of flying conditions.

When I first started learning how to land a Pitts S-2B, I found myself having a tendency to keep one wing low during the approach to allow a better view of the runway while flying in little or no crosswind. This resulted in my touching down on one main gear while the other was still off the ground. After I noticed I was doing this, I began to force myself to keep the wings level during the approach, and I then became accustomed to the "picture" of how the plane and runway should look while landing. This is only one example of how pilots can get into the habit of not landing with the wings level. As you land, critique yourself and develop a feel for the plane. After some practice, you will find that it becomes much easier to sense what the plane is doing during takeoffs and landings, and you will be able to compensate with ease. In the following chapters we will cover other takeoff and landing situations, but keep in mind the basics we covered in this chapter.

6
Short-field takeoffs and landings

IN THIS CHAPTER WE ARE GOING TO DISCUSS THE FUNDAMENTALS OF short-field takeoffs and landings. There's more to making a short-field takeoff than pushing the throttle to the fire wall and hoping you can clear the trees at the end of the runway. Understanding the limits of the plane for takeoff goes a very long way in helping you prevent an accident. The same is true when it comes to short-field landings. While the objective is to get down and stopped in the shortest possible distance, not knowing what the plane is capable of, or not using the correct techniques during the landing, can result in bruised pride, bent airplanes, or worse.

We will begin our explanation of short-field landings with a review of the various types of takeoff and landing charts. These charts are for imaginary planes, but the fundamentals for their use will hold true for actual aircraft. We will then move on to takeoffs, covering flaps and trim settings as they are related to short-field takeoffs. Then we will cover techniques for ground roll, rotation, and climb for both conventional-gear and tricycle-gear planes. Short-field landings follow, with flap and trim settings reviewed first, followed by approach, flare, touchdown, and rollout procedures. These will again be for both taildraggers and tricycle-gear planes.

We will cover commonly made mistakes for both takeoffs and landings. I will also include personal experiences related to these situations. When you finish the chapter, you should have a solid understanding of the correct techniques to use. As stated frequently in this book, these techniques are general in nature and have worked well for me. If the manufacturer of the plane you fly recommends other procedures, follow those. Short-field takeoffs and landings assume a maximum-performance situation, so many of the concepts we have previously covered (such as airspeed control, glideslope control, flap use, and pattern procedures) become extremely important. You will be flying the plane to the edge of its safe limits, and if you have not performed short-field takeoffs and landings for some time, or if you feel uncomfortable with any portion of the procedures, find a qualified flight instructor to help you out. Mistakes made during the approach for a short-field landing, such as too high an airspeed or too shallow a glideslope, can result in additional runway length being used for takeoffs and landings.

PERFORMANCE CHARTS

In this section we will look at examples of takeoff and landing performance charts for imaginary aircraft. Charts are used to determine the length of the runway you will require for takeoffs and landings. These charts are developed by the aircraft manufacturer and can be found in the operations manual of the plane you fly. I highly recommend using them anytime you are flying in conditions for which you have not previously computed safe landing and takeoff requirements, such as changes in temperature, pressure altitude, unfamiliar runways, or changing gross weights. All of these factors affect the distance you will need to take off or land. You should also use performance charts anytime you are flying an airplane that is new to you.

The charts in this chapter are examples of what you might find as you are reviewing charts for various aircraft and are not intended to represent the charts for a particular make or model of aircraft. After you finish the section, you should have an understanding of how to compute the takeoff or landing distance of an airplane. When you calculate the distances for the plane you are flying, follow the manufacturer's recommended procedures. They might differ somewhat from the examples in this chapter, but you will understand the basic concepts necessary to accurately compute takeoff and landing distances.

Before we begin looking at the charts, I want to define two terms you will see on them. The first is *pressure altitude*. This is the altitude the altimeter shows when the pressure in the barometer "window" is set to 29.92. This is the altitude in the standard atmosphere corresponding to a particular pressure level. The second is *density altitude*. Density altitude is the pressure altitude corrected for nonstandard temperature. When temperature for a particular altitude matches the "standard" values, the pressure altitude and density altitude will have the same value. Temperature and pressure affect density altitude. For instance, when temperatures are above the norm for a standard day, the density altitude will be higher than the pressure altitude. If temperatures are lower than the standard value, the density altitude will be lower.

Density altitude directly affects how the airplane will perform. The amount of thrust the engine produces and the amount of lift generated by the wings are affected by both pressure altitude and density altitude.

The plane's analog instruments, such as the altimeter, sense changes in pressure and react to those changes. So does the airplane itself. Instead of moving a needle to display an altitude value, the airplane changes its performance characteristics as related to takeoff and landing distance requirements. Performance charts give you insight into exactly how the plane will respond to a given set of conditions.

TAKEOFF CHARTS

Figure 6-1 is the first table we will use in this takeoff calculation example. Note that it has ten columns. The first column is composed of several different gross weights for the plane, and column two represents the headwind component. These two columns are paired together with multiple headwind factors for each gross weight. The following eight columns are also arranged in pairs. Each pair has columns indicating ground run distance and distance to clear a 50-foot obstacle. Each of these pairs has a heading that indicates the pressure altitude the plane is operating at and the standard temperature for that altitude. The top of the chart indicates that the distances computed are when the plane is flown from a hard-surface runway with 20° of flaps. At the bottom of the chart, a note states that you should increase the takeoff distance 10% for each 25°F above the standard temperature for that altitude.

Let's briefly discuss standard temperature and how it relates to aircraft performance. Most aircraft performance charts use a "standard day" set of conditions for the base calculations in the chart. A standard day is defined as sea-level altitude, a temperature of 59°F, and an atmospheric pressure of 29.92 inches of mercury. As the chart

Hard-surface runway/Flaps at 20°

Gross weight	Head-wind	Sea level/59 F		2500 Ft./50 F		5000 Ft./41 F		7500/32 F	
		Ground run	50-Ft. obstacle	Ground run	50-Ft. obstacle	Ground run	50-Ft. obstacle	Ground run	50-Ft. obstacle
	0	723	1431	987	1832	1207	2340	1529	3791
2400	15	502	998	696	1315	874	1889	1032	2812
	30	115	350	200	456	263	533	328	740
	0	789	1501	1035	2015	1311	2590	1600	3909
2650	15	556	1112	725	1476	921	2000	1231	3010
	30	156	379	223	498	292	588	361	814
	0	867	1654	1139	2217	1456	2898	1762	4302
2900	15	606	1268	803	1611	1031	2205	1388	3320
	30	191	411	253	548	327	644	399	902

**Increase distances 10% for every 25 F above standard temperature.

Fig. 6-1. *Takeoff-distance chart.*

shows, the standard temperature goes down as the altitude becomes higher. I still do not know why this particular set of values was originally chosen. In more than 20 years of flying, I don't think that I have flown on a standard-day set of conditions. When conditions vary from the standard-day values, they affect the performance from the plane, and these differences must be accounted for when you do the calculations. The note at the bottom of the chart shows how the takeoff distance increases 10% for every 25° increase above standard temperature for an altitude. It should become instantly clear why you use less than half the runway taking off on a cold, crisp day in December, and you use more than three-quarters of the runway on a hot, humid day in August. The pressure altitude is higher, and the temperature in the summer might be 80° or 100° higher than during the winter. The chart graphically depicts how large the difference in runway requirement for takeoff can be!

At this point we will use the chart to determine takeoff requirements for two examples. For this chart there are four steps to computing the distance:

1. Find the gross weight under the gross-weight column.
2. Find the headwind under the headwind column in the same row as the gross weight.
3. Follow the headwind row right to the first column for the altitude that corresponds to your takeoff altitude. The value at the intersection of this row and column is the length of the ground run or distance to clear a 50-foot obstacle in feet for that set of conditions. This is if the temperature is standard for that altitude.
4. Increase the value found in step three by 10% for each 25° the temperature is greater than the standard temperature for that altitude. The resulting value is the ground-run or 50-foot-obstacle clearance distance.

Example 1

Determine the ground run and distance to clear a 50-foot obstacle for the following conditions:

- Gross weight: 2400 lbs
- Pressure altitude: sea level
- Temperature: 84°F
- Headwind: 30 mph

Start by finding the correct gross weight row in the first column—2400 lbs. Move right to the second column next to that gross weight and find the corresponding headwind, 30 mph, then move right along the same row for the headwind value to the sea-level column. We can see that the ground run is listed as 115 feet and the distance to clear a 50-foot obstacle is 350 feet. Since the temperature is 84° and is 25° above the standard temperature, we will need to add 10% to the values to compensate. This means 115 + 12 = 127 feet for the ground roll and 350 + 35 = 385 feet for the distance to clear the 50-foot obstacle.

Example 2

Determine the ground run and distance required to clear a 50-foot obstacle for the following conditions:

- Gross weight: 2650 lbs
- Pressure altitude: 7500 ft
- Temperature: 82°F
- Headwind: 15 mph

In this example we start at the 2650-pounds row in the first column, move right to the 15-mph row in the next column, then follow that row right until we are under the 7500-foot heading. The initial ground run is 1231 feet, and the distance to clear a 50-foot obstacle is 3010 feet. Because our temperature is 50° above the standard temperature, we will need to increase the distances by 20%. As a result, 1231 + 246 = 1477 feet for the ground run, and 3010 + 602 = 3612 feet to clear the 50-foot obstacle. As you can see, there is a significant increase in the distance required between sea level and 7500-foot pressure altitudes for any gross weight or headwind condition. These types of charts might have additional notations that can increase or decrease the final value for a given set of conditions. For instance, if you fly from a grass runway instead of a hard-surface runway, the chart might indicate you should increase the distance by a certain percentage, while other charts might indicate that flaps should be in the raised position for takeoff. Some I have seen do not have a headwind column but instruct you to reduce the values by a given percent for each multiple of a given headwind component. Follow the instructions for your airplane, and use the procedures recommended by the manufacturer.

Figure 6-2 is an example of another type of takeoff distance chart. This chart makes use of graphs as opposed to a table. Let's walk through the procedure for determining takeoff distance with this chart. At the top of the chart you will note that only one gross weight is indicated for the plane. It also indicates that the values are valid for takeoff on a paved, level, dry runway, and that full power should be applied prior to release of the brakes. Finally, the values are for no-wind conditions and no flaps. There are two lines on the graph, one marked ground roll and the other for over a 50-foot obstacle. The left side of the graph indicates the density altitude, and the lower portion of the graph is the takeoff distance.

To find the takeoff distance for the conditions outlined at the top of the chart, you complete the following steps:

1. On the left side of the chart, find the position of the density altitude corresponding to the airport you are taking off at.

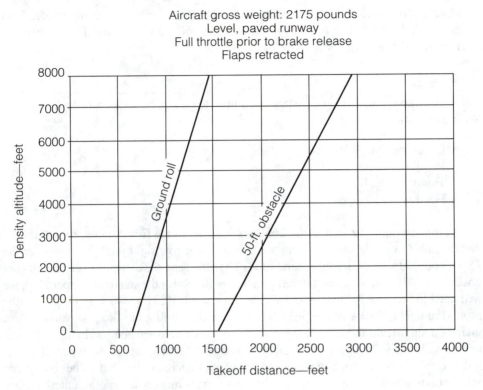

Fig. 6-2. *Takeoff-distance chart.*

2. Move horizontally from that point to the right until you intersect the ground-roll line or the 50-foot-obstacle line.

3. From the point of intersection, move down vertically to the bottom of the graph.

The corresponding value will be for the ground-roll or takeoff distance over a 50-foot obstacle.

Example 1

Find the ground roll and distance to takeoff over a 50-foot obstacle for the following conditions:

- Density altitude: 1500 ft
- Gross weight: 2175 lbs
- Headwind: 0 mph

In this case we start on the left side of the graph and find the midway point between 1000- and 2000-foot density altitude. We move right until we intersect the ground-roll line. To find the ground-roll distance, we then move down vertically to the bottom of the graph. Here we see the value is approximately 850 feet. To determine the distance to clear a 50-foot obstacle, we would have followed our horizontal line for 1500 feet past the ground-roll line until we intersect the 50-foot-obstacle line. We then move down vertically to the bottom of the graph to find that the distance is approximately 1875 feet.

Example 2

Find the ground roll and distance to clear a 50-foot obstacle for the following conditions:

- Density altitude: 5000 ft
- Gross weight: 2175 lbs
- Headwind: 0 mph

In this example we start on the left side of the graph at the 5000-foot density altitude value and move right horizontally until we intersect the ground-roll line.

At the point of intersection, we move down vertically to find that the ground-roll value is approximately 1200 feet. The 50-foot-obstacle-clearance value is found by following the 5000-foot line right until it intersects with the 50-foot-obstacle line, and again moving down vertically to the bottom of the chart. In this case the distance to clear a 50-foot obstacle is approximately 2625 feet.

This particular type of chart is accurate for only a specific set of conditions pertaining to gross weight, wind, and other factors. You can see by comparison to the column chart we just looked at that there is not as much flexibility in determining takeoff distances. Like the column chart, the graph chart might also have notations stating that differences from the noted conditions, like grass runways or headwind values, can reduce or increase the computed value by a given percentage. Again, this chart is not for a specific airplane and is used only for the purpose of explaining how this type of chart might be used. Follow the instructions laid out by the manufacturer of the plane you are flying. Now let's move on to landing-distance charts.

LANDING CHARTS

In this section we are going to look at landing-distance charts and how to use them. The chart formats will be similar to the ones we have used in our takeoff-distance examples. Like the takeoff charts, these are only examples and are not intended for use with a specific airplane you might be flying.

Figure 6-3 is composed of columns much like the takeoff-distance chart in Fig. 6-1. The headings consist of gross weight and airspeed on final, followed by ground-roll and distance-to-clear-a-50-foot-obstacle column, paired under four different alti-

Hard-surface runway/Flaps at 40°/Power off

		Sea level/59 F		2500 Ft./50 F		5000 Ft./41 F		7500/32 F	
Gross weight	Airspeed on final	Ground roll	50-Ft. obstacle	Ground roll	50-Ft. obstacle	Ground roll	50-Ft. obstacle	Ground roll	50-Ft. obstacle
2400	66	502	998	696	1315	874	1889	1032	2812
2650	69	523	1112	725	1476	921	2000	1231	3010
2900	72	606	1268	803	1611	1031	2205	1388	3320

**Decrease distances 10% for every 6 mph of headwind.

Fig. 6-3. *Landing-distance chart.*

tude headings ranging from sea level to 7500 feet. Notes on the chart indicate you should reduce landing distances by 10% for each 6 mph of headwind, and the distances are for power off and 40° of flaps. To determine the landing distance for a given set of circumstances, you will perform the following steps:

1. Select the gross weight of the aircraft under the gross-weight column.
2. Move right to the approach-speed column, and it tells you what the correct approach speed should be for that gross weight.
3. Continue to move right until you are under the correct-altitude column.
4. Under that column is the ground roll needed for landing and the distance for landing required to clear a 50-foot obstacle.

If you have a headwind, you will reduce the distance required for landing by 10% for each 6 mph of headwind component. Now let's step through two examples using the chart.

Example 1

Determine the ground roll and distance to clear a 50-foot obstacle for the following conditions:

- Gross weight: 2650 lbs
- Pressure altitude: sea level
- Temperature: 59°F
- Headwind: 18 mph

We start in the gross weight column and find the 2650-pound row. Moving right, we see that the correct approach speed is 69 mph for this gross weight. We continue to move right until we are under the sea-level column pair. Here we find that the ground-roll value is 523 feet, and the distance to clear a 50-foot obstacle and land is 1112 feet. These values are for calm conditions, and in this case we have an 18-mph headwind component. This means we will reduce the landing distance by 30%. As a result, the ground roll becomes 523 – 157 = 366 feet, and the distance to land over a 50-foot obstacle is reduced to 1112 – 334 = 778 feet. Strong headwinds can be very helpful in making short-field landings very short. (They can make it more difficult to try to glide to a runway when you lose your engine. More about my experiences with that in chapter 11 on emergency landings, though.)

Example 2

Determine the ground roll and distance to clear a 50-foot obstacle for the following conditions:

- Gross weight: 2400 lbs
- Pressure altitude: 7500 ft
- Temperature: 32°F
- Headwind: 6 mph

We again start in the gross-weight column, and this time find the 2400-pound row. Moving right, we see that the correct approach speed is 66 mph for this gross weight. We continue to move right until we are under the 7500-foot-altitude column pair. The ground roll value at this altitude is 1032 feet, and the distance to clear a 50-foot obstacle and land is 2812 feet. These values are again for calm conditions, and in this case we find we have a 6-mph headwind component. This means we will reduce the landing distance by 10%. As a result, the ground roll becomes 1032 – 103 = 929 feet, and the distance to land over a 50-foot obstacle is reduced to 2812 – 281 = 2531 feet.

Next we will cover a graph-type chart for landing distances. Figure 6-4 will be used in this example and, like its takeoff-distance counterpart, has a number of notations associated with it. It again specifies the gross weight for the aircraft, that the runway needs to be paved, level, and dry, that the landing is power off, and 40° of flaps are used. The chart, like the table graph, specifies the approach speed as well.

To use the graph, perform the following steps:

1. Find the correct density altitude on the left side of the graph.
2. Move right until you intersect either the ground roll or distance-to-clear-a-50-foot-obstacle line on the graph.
3. Move down vertically from the point of intersection with the appropriate line to the bottom of the graph. The value at the bottom of the graph is the corresponding ground roll or distance to land over a 50-foot obstacle.

Let's look at two examples for this type of landing chart.

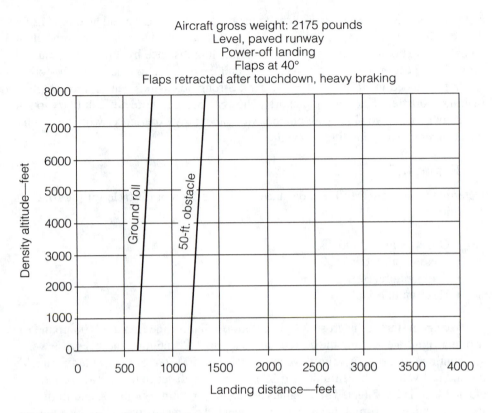

Fig. 6-4. *Landing-distance chart.*

Example 1

Find the ground roll and distance to clear a 50-foot obstacle for the following conditions:

- Density altitude: 5000 ft
- Gross weight: 2250 lbs
- Headwind: 0 mph

We start on the left side of the chart at the 5000-foot density-altitude value. From there we move right until we intersect the ground-roll line. Moving vertically down to the bottom of the graph, we find the ground-roll value to be approximately 725 feet. To determine the landing distance over a 50-foot obstacle, we would find the intersection point between the 5000-foot line and the 50-foot-obstacle line, then move down vertically to the bottom of the graph. In this case the distance is approximately 1295 feet. You can see that the results are not extremely accurate and are open to some interpretation, depending on how much detail the values of the graph contain. Someone else might look at the same situation and feel it reads 1290 feet.

Example 2

Find the ground roll and distance to clear a 50-foot obstacle for the following conditions:

- Density altitude: sea level
- Gross weight: 2250 lbs
- Headwind: 0 mph

In this example we start on the left side of the chart at the 0-foot density-altitude value. Again we move right until we intersect the ground-roll line. The point of intersection between the 0-foot line and the ground-roll line indicates that the ground-roll value is approximately 630 feet. To once again determine the landing distance over the proverbial 50-foot obstacle, we find the intersection point between the 0-foot line and the 50-foot obstacle line. In this case we find the distance is approximately 1190 feet. Since we are at the bottom of the graph, as we move across the 0-foot line, there is no moving down vertically. All we need to do is find the intersection point between the bottom of the graph and the lines for ground roll and distance to clear a 50-foot obstacle.

There are many types of takeoff and landing-distance charts and graphs that can be included in the operations manual for an airplane. Some graph charts have multiple sections, with each section determining a different factor such as altitude, headwind component, and landing speed. These graphs are fairly complex, but they offer accurate landing and takeoff information for varying conditions. Be sure to use the chart the manufacturer supplies for your airplane. Even charts for different models from the same manufacturer might vary a great deal in what the results will be. If you are not sure how to use the chart for your plane, ask a qualified flight instructor to work through some examples with you. In any situation, but in particular short-field work, you should never assume the runway is long enough. Now that we have looked at how to use charts, we will discuss short-field takeoff and landing techniques.

TAKEOFF

In this section we will review short-field takeoff techniques for both tricycle- and conventional-gear aircraft. When you have completed this section, you should understand the basic techniques and practices to use when making short-field takeoffs. Included will be examples of things to avoid during short-field takeoffs as well. Like the previous chapter, the format will include trim settings, flaps settings, and ground roll and rotation topics.

Flap settings

The basic thing to remember concerning flap settings is to adhere to the manufacturer's recommendations. Some manufacturers recommend no use of flaps. Others recommend relatively low flap settings. The use of 10° of flaps might reduce the ground run, but it might also increase the distance necessary to clear an obstacle. On other aircraft, the use of flaps might reduce both the ground run and distance to clear an obstacle. On some aircraft use of flaps might increase all distances due to the additional drag flaps create. Your airplane's operations manual will indicate the correct procedure to follow, so be sure to read the manual and become familiar with it.

If flaps are recommended, be sure to have sufficient airspeed and altitude prior to retracting them after leaving the ground or clearing the obstacle. As we have previously noted, the plane might have a tendency to settle as the flaps are retracted, or, if you are too slow, the plane might also stall. Keep your airspeed at an acceptable value prior to flap retraction, and transition to the appropriate climb airspeed with the flaps retracted. Sufficient altitude at the time of retraction will allow for some settling without contacting the ground. The tendency to settle might be more pronounced at slower airspeeds or with higher gross weights. Retract the flaps slowly. In many cases, it is a good idea between flap increments to allow the plane to stabilize prior to the next increment of retraction. In most cases, it is bad practice to slam the flaps into their retracted position. This is not normally a concern with electric or hydraulic flaps, but it can happen with manual flaps.

It is also not usually recommended to add flaps for a boost in lift after the plane accelerates without flaps. While on the surface this might seem like a good idea, the introduction of the pitching moment created by adding the flaps might cause the plane to either pitch down or up. These are equally large problems when you are close to the ground.

Trim settings

Trim is usually set to normal values for short-field takeoffs. You should use the standard setting recommended for your aircraft. If you anticipate the steeper climb angle used for a short-field takeoff and add slightly more nose-up trim to help reduce pitch-control pressures, you might find the plane leaving the ground before it is ready to fly. Trim the plane for normal takeoff and fly the plane. After you are off the ground and have established the correct climb speed, you can then retrim the plane for lighter pitch pressures. Depending on the length of time you must stay at the recommended short-field climb speed before transitioning to the recommended cruise-climb or best-rate-of-climb speeds, it might not be necessary to reset the trim. Once the plane is off the ground, the obstacle might be cleared in only a few seconds, and you might not find it necessary to retrim. Attempting to divide your attention between flying the short-field takeoff and setting the trim at a low altitude might not be a wise procedure. You will have to decide what you are comfortable with, but keep safety as the primary task during your takeoff.

Conventional-gear ground roll and rotation

In this section we cover the short-field takeoff procedures for a conventional-gear airplane. The purpose of a short-field takeoff is to become airborne in as short a distance as possible and, when necessary, to safely clear the ever-present 50-foot obstacle at the end of the runway. I have had to take off from a number of short strips, but only on rare occasions have I had a serious obstacle to clear. You will have a greater need for clearing obstacles on a regular basis if you do more bush flying, where trees nudge right up to the end of some runways, or if you do more flying from the center of a city, where tall buildings are sometimes located close to the end of the runway.

In our discussions we will again assume either no wind or a direct headwind. Crosswind techniques will not be discussed until chapter 8.

Our review will center around general short-field takeoff techniques, and the review is not intended to represent a particular aircraft. If your aircraft's manufacturer recommends that you use a different technique for short-field takeoffs, use it. One characteristic of competent, proficient pilots is the ability to adapt to different planes, techniques, and conditions. Use the information we cover in this chapter to enhance your knowledge and abilities, not as the only way to perform a short-field takeoff.

As a rule you will become airborne on short-field takeoffs at a speed that is slower than that of a normal takeoff for that plane. Because of this, you will have reduced control effectiveness, and the plane might feel somewhat "mushy" in how it responds to your control inputs. As we discuss the takeoff roll, rotation, and climb, we will touch on points to keep in mind related to control effectiveness.

To begin, a short-field takeoff should allow you to use all of the runway available. As you taxi onto the runway, end up with the tailwheel as close to the end of the runway as possible. Be sure to get the tailwheel centered or locked, depending on your tailwheel design, prior to beginning your takeoff run. Full throttle application is not the time to find out that your tailwheel is turned 90°! It might be advisable to let the tailwheel leave the end of the runway as you make your turn to align with the centerline, and then center it as you taxi forward and bring it on to the runway. Only attempt to do this if the surface conditions allow for it. Soft ground or snow might cause you to become stuck, or a steep drop from the runway to the ground might damage your tailwheel, so use your judgment before trying this.

A short-field takeoff is accomplished by raising the tailwheel from the ground, but to a lesser degree than with a normal takeoff. In most cases, 4 to 8 inches is the range you want to have the tailwheel off the ground. This might vary, though, depending on the type of plane you are flying. The purpose is to maintain the plane's wings at a higher angle of attack than during a normal takeoff. This helps to generate more lift and allows the plane to become airborne at a slower airspeed. Most aircraft-operations manuals indicate the airspeed the plane should leave the ground at, so use that as the target airspeed. If you have the tailwheel too low, the plane might take off at too slow an airspeed and stall, or settle back to the runway after it leaves ground effect. If the angle of attack is too low, the plane will require additional airspeed, and runway, to become airborne.

There are two schools of thought related to throttle advancement with short-field takeoffs. One method calls for locking the brakes and advancing the throttle to full power, then releasing the brakes. However, with tailwheel aircraft there is a danger of nosing the plane over as you advance the power. One technique recommended to overcome this nosing-over tendency is to maintain full back-control stick, forcing the tail on the ground. If you do not have the stick all the way back, or have a very high-horse-power engine, you might still have a problem with nosing over, though, so it is necessary to know your airplane before attempting this. The rationale for doing this is that the engine has time to run up to full power before the plane begins to move, and this will result in a shorter takeoff run.

The other technique is to advance the throttle without holding the brakes. This technique avoids the tendency to nose over in a tailwheel plane. I have read several articles stating that piston engines develop full power very quickly, and locking the brakes results in no measurable reduction in takeoff distance. The practice of locking brakes does make a difference with turbine engines, which can take a considerable amount of time to spool up to full power. I must confess that I have not been able to determine any difference in short-field takeoff distances with either technique. You might have had different experience depending on the plane you fly, so use the technique that works best for your aircraft.

After the plane begins to roll, you will need to input a slight amount of forward stick. This will be less than the amount to raise the plane to the normal takeoff attitude, so it might require practice to get the tail off the ground in the four- to eight-inch range. After getting used to reaching a slightly tail-low or tail-level attitude for takeoff, you will need to adjust your elevator input to get the plane at the correct attitude for a short-field takeoff. Due to the lower takeoff speed, the elevator might be less effective, so your inputs might need more travel to control the pitch. You might find it also takes more right rudder travel to counteract the left-turning tendencies we have previously discussed. This in turn would require more aileron input to maintain level, coordinated flight.

As the plane leaves the ground, you should transition to a climb at the recommended airspeed. If you need to clear an obstacle, this will likely be the best-angle-of-climb speed, or V_x. As you recall, this allows the plane to gain the largest amount of altitude in the shortest distance. After the obstacle is cleared, you should transition to V_y, the best-rate-of-climb, or the recommended-cruise-climb speed. These airspeeds allow for better engine cooling. Due to engine overheating, V_x should not be used for prolonged periods.

At a nearby airport, the runway is only 21 feet wide, and the runway is virtually impossible to see when the tail of the S2B is not up in the normal takeoff stance. When making short-field takeoffs from there, I can see only a very small part of the runway in my peripheral vision. On a narrow runway like this, it becomes extremely important to start out heading straight down the center and maintain the correct track as you move forward. Prior to turning onto the runway, you should look down its length to be sure it is clear as you make the takeoff roll. During the takeoff roll on narrow runways, there is much less room for error, so you need to be precise in your piloting and correct immediately for any deviation from the runway centerline. Depending on the left-turning tendency your plane exhibits and the length of the runway, you might want to feed power in a little slower than usual to aid in keeping the plane straight as the takeoff roll begins.

Any time you are making a short-field takeoff, you should also determine what the abort factors will be. Most important among them is an abort point along the length of the runway. If you have not become airborne when you hit that point on the runway, do not continue the takeoff. This abort point is, to a certain degree, going to depend on your comfort level with the takeoff scenario. If you are making a short-field takeoff that will use most of the runway, you might not have sufficient runway length to brake the plane to a stop after accelerating to almost takeoff speed. In these situations the takeoff charts and accurate weather information will help you calculate the amount of runway you will need. If the margin of excess runway is small, you will need to evaluate to determine your level of confidence with the situation. In these narrow-takeoff-margin situations, you will need to be evaluating the takeoff as it progresses to determine if it should be aborted.

As you begin the takeoff roll, scan the engine instruments to verify that the engine is producing full power. If oil pressure, fuel pressure, RPMs, or manifold pressure are low, you should abort immediately. As the aircraft is accelerating, "feel" the plane to see if it is the normal acceleration rate you have become accustomed to. If the acceleration seems slow, this is another reason to abort the takeoff. Do not force the airplane off the ground in an attempt to get off a short runway. There have been cases of pilots getting the plane off the runway, but then they are unable to accelerate and climb out of ground effect, crashing somewhere past the runway.

You should also have a contingency plan in mind before you begin the takeoff run if you need to clear an obstacle but find you can't. Depending on the terrain, it might be possible to turn to avoid the obstacle. There is no answer that will work in all situations of this type, but keep your options open by thinking of them ahead of time. Twenty-five feet off the ground is no time to begin to think about how you are going to avoid the grove of trees that you are not going to be able to clear.

A commonly exhibited problem with short-field takeoffs is not raising the tail the correct amount from the runway. If the tail is too high, the angle of attack of the wings will be less than the correct amount, and it will take more runway and airspeed to achieve the lift necessary to lift off. When the tail is too low, you will generate more drag as a result of the excessive angle of attack. This additional drag might result in a longer takeoff run as well. The excessive angle of attack might allow the plane to lift off at too slow an airspeed, which, as we have previously discussed, could result in a stall as the airplane climbs. Another common error you might encounter is uncoordinated flight. The higher angle of attack will require additional right rudder and left ailerons to fly in a coordinated manner. Finally, not using all available runway is a very common problem. I have had students taxi onto the runway and go 30 to 50 feet down it before they get lined up and ready for takeoff. This extra distance could make a difference on a hot day at high gross weight. Use all the runway and all the power the plane has.

If you have not performed short-field takeoffs in a taildragger, you should get instruction from a qualified flight instructor. Short-field takeoffs can be performed as safely in a taildragger as in a normal takeoff, but it takes practice and knowing the limits of the plane you fly. In the next section, we will discuss short-field takeoff techniques for tricycle-gear planes.

Tricycle-gear ground roll/rotation

In this section we will review the correct procedures for the takeoff roll and rotation in a tricycle-gear aircraft. Like the techniques for conventional-gear planes, we will cover general techniques that I have found to work well for me. If your plane's operations manual indicates the use of a different set of procedures, follow those. When you have completed this section, you should understand the short-field takeoff techniques to use and the problems that are commonly found.

In many ways, a tricycle-gear, short-field takeoff is similar to the conventional one we just covered. You want to get the plane into the air safely in the shortest distance possible. Most aircraft operations manuals I have reviewed indicate that you should leave the nosewheel on the ground until a certain speed is reached, then rotate and fly away. There have been some manuals, though, that indicate you should begin the takeoff roll with the yoke back to put the plane in a slightly tail-low position. Those manuals that indicate the nosewheel should remain on the ground until rotation also indicate that raising the nosewheel early will increase the angle of attack and resulting drag, resulting in a longer ground run. Each airplane is different, and you should become familiar with its characteristics and flight manual's recommendations. Never assume all planes use the same techniques.

The short-field takeoff begins with positioning the plane as close to the end of the runway as possible. Without a tailwheel, it is possible to put the main landing gear on the edge of the runway with the tail hanging out over it. Be careful that no runway lights or other obstacles are positioned in such a way that the tail might strike them, though. You can use either of the throttle advancing techniques we covered for taildraggers. Hold the brakes and advance to full throttle before releasing them, or advance to full throttle without holding brakes. If you choose the first option, be sure to hold between neutral to full-back elevator to prevent the plane from nosing over. If you hold full-back elevator, remember to go to a neutral to slightly tail-low elevator input after you have released the brakes. Maintaining full-aft elevator input after the brakes are released could result in the nosewheel popping off the runway well below the safe takeoff airspeed.

As you reach the recommended rotation speed, smoothly feed in up-elevator input to raise the nose of the plane to the correct climb attitude. Knowing the correct attitude will come with practice and will allow a smooth transition to the climb. The correct climb speed will depend on whether you are attempting to clear an obstacle at the end of the runway or can enter a standard climb at V_y or the cruise-climb airspeeds.

You will be leaving the ground at a slower airspeed than with a standard takeoff and, like the conventional-gear plane, you will find reduced control effectiveness until the plane accelerates. This will require the use of additional right rudder, left aileron, and possibly elevator to control the plane. Be sure not to overcontrol at this slower airspeed and low altitude. Although our examples are for nonheadwind conditions, turbulence could cause the plane to move around, and you might feel the plane is close to a loss of control because of how "mushy" the controls feel. On gusty or turbulent days, it might be necessary to lift off at a higher airspeed to maintain control of the plane. Obviously this will require more runway during the takeoff, but flying at too slow an airspeed can compromise safety. You will need to judge the situation to determine the best course of action. Remember that one option is to wait until conditions permit safe flight.

Common mistakes include not using the entire runway for takeoff, incorrect use of flaps, incorrect pitch angle during the takeoff roll, and incorrect rotation and climb speed. Each of these errors can be eliminated by practicing short-field takeoff and knowing the recommended procedures and airspeed in the plane's operations manual. You are flying the plane near its limits, and even minor mistakes in piloting technique will increase the length of the takeoff run. Never assume as you move from one plane to another that, even though they are from the same manufacturer, the short-field takeoff procedures are the same for all of them.

LANDING

In the following sections we will discuss short-field landing procedures and techniques. Like short-field takeoffs, the idea is to use the shortest amount of runway during the landing. To consistently accomplish good short-field landings, you will need to use all of the techniques we have discussed in previous chapters. Picking your landing spot and recognizing whether you will be short or long in hitting it becomes extremely important. Maintaining a constant and correct airspeed and glideslope are equally important in your short-field approach, as is the correct use of power. Without having these principles down, it will be difficult to perform good short-field landings time after time. Even if the charts tell you it is possible to get into a given runway, you might find it is too short if you touch down long or carry too much airspeed on final. By the same token, dragging the plane in might cause you to touch down short of the runway, or worse yet, touch down on the 50-foot obstacle you are trying to clear.

I have seen more than one student display a certain amount of concern and anticipation at the prospect of attempting to land on an actual short field. Simulated approaches on a runway that is 10,000 feet long and 150 feet wide will help you learn the basics, but take a qualified instructor with you and put your practice to work on an actual short runway. There is a real feeling of accomplishment to know you can get the plane down safely by flying it within the manufacturer's recommended procedures. In this section we will cover some of the techniques you can use. As always, this section will include both conventional-gear planes and tricycle-gear aircraft. Let's begin with a review of flap settings.

Flap settings

The correct use of flaps, if your plane is equipped with them, can play an important part in a well-executed short-field landing. As we have already seen, flaps allow the plane to land at a slower airspeed and maintain a steeper angle of descent. The slower landing speed results in less runway needed to stop the plane. The steeper glideslope allows you to clear the obstacle at the end of the runway and have less distance between the obstacle and the point at which you will touch down.

Most operations manuals I have seen recommend the use of full flaps for short-field landings. Assuming you have an airplane with three notches of flaps, I like to put the first notch on after I have turned on base, the second notch on after the initial turn to final, and the third notch on while on short final.

Keep in mind our discussion in previous chapters concerning the need to compensate to maintain the correct airspeed and glideslope as you add each flap increment. This becomes especially important in short-field landings, where the airspeeds are slower and the glideslope steeper. When you are operating at 10 or less knots above stall speed, you must hold the target airspeed very accurately or you might end up as one of those "stalled on final without sufficient altitude to recover" statistics. As you get used to the airplane you fly, you should develop a "feel" for changes in airspeed. The noise of the engine, propeller, and wind stream over the plane all will be used to indicate when the plane is gaining or losing airspeed. Control effectiveness will also be an indication of airspeed changes. These sensory inputs can be extremely informative if you become attentive to them. This will allow you to keep your eyes looking outside the cockpit looking at the runway, pattern, and other areas of interest as you approach to land and spend less time looking at instruments inside the cockpit. I know a pilot who practices landings at night with the panel lights shut off. He wanted to learn to "listen" to the airplane and not use the instruments as a crutch when they were not necessary. I recommend that you do not attempt this particular test, but it illustrates how this pilot could judge his airspeed by the sounds of the wind and the feel of the controls. Short-field approaches are not the time to keep your eyes inside the plane. You should be looking and compensating to assure that you land on your target touchdown point.

Flap extension will affect the "feel" of the plane. There will be a tendency for the nose to pitch, the plane to slow down, the glideslope to increase, and the controls to become less effective. As you make the approach, you should be looking at your intended touchdown point and determining if your glideslope will take you to it. For each notch of flaps you add during the approach, you might need to alter another control input to maintain the correct glideslope to the touchdown point. When adding flaps, you might need to increase or decrease power or adjust your pitch. As you come to know your airplane better, you will be able to anticipate your plane's reaction to each additional notch of flaps.

Trim settings

During a short-field approach, trim is adjusted in the same manner as during a normal approach. As a result of the slower airspeeds without trim adjustment, you might need to maintain more back pressure with the elevator to hold the correct airspeed. Without resetting the trim adjustments, momentary lapses in attention to your elevator input could result in an unwanted increase or decrease in airspeed as the nose pitches up or down. As a result, you would have a less precise approach.

Add trim in the same method we discussed in previous chapters. After you pull the power on downwind, maintain your altitude and let the airspeed bleed off to the desired value. Then add trim to bring the elevator back pressure input to a neutral feel on the control yoke. As you add each notch of flaps, you might find it necessary to retrim during the approach to maintain the correct elevator input feel.

One brief note of caution on a topic we will cover in greater detail in chapter 11. If you find it necessary to abort the landing, you will find that with a great deal of nose-up trim, such as you might use in a short-field approach, as you apply full power to go around, the plane will want to pitch up dramatically. This is due, in part, to the amount of nose-up trim you have added during the approach. Be prepared for this pitching moment and the amount of forward pressure it might require to maintain the correct airspeed during the climb out. You will not damage the plane by preventing the nose from rising above the correct pitch for the climb. As you get the situation under control, reset the trim to remove any unwanted pitching moment. You must make the airplane do what you want it to. Don't let it lead you into a potentially dangerous situation.

Conventional-gear flare and landing rollout

In this section we are going to look at the techniques for a safe, accurate short-field landing. A good, solid, consistent approach is the best way to land in the shortest possible distance, so start flying the approach accurately from the very beginning, while you are still on downwind. During your final approach, monitor whether you are going to land long or short of your intended touchdown point, and adjust your glideslope as necessary using techniques described in chapter 4.

If you have an obstacle to clear on final, there are two basic methods you can use to clear it and touch down on the runway. The first is to fly a flatter approach with a certain amount of power. Then, after you clear the obstacle, reduce your power to increase the glideslope for the rest of the approach to landing. When you execute this type of approach, you will need to monitor your airspeed and be prepared for the increased rate of descent as you reduce the power setting. Figure 6-5 illustrates the change in glideslope before and after the obstacle is cleared. The second method calls for setting up and maintaining a steeper glideslope from the initial final approach until touchdown. This steeper glideslope is initiated from a higher altitude and provides adequate clearance over the obstacle. Figure 6-6 is an example of this steeper approach over the obstacle. In this method, power remains constant, as does the angle of the glideslope and airspeed.

There are situations where one method might be more appropriate than others. For instance, in controlled airspace, ATC might prevent you from climbing to an altitude that will allow you to set up the second approach method, so you will need to carry power until the obstacle is cleared, then reduce it. The obstacle you are trying to clear

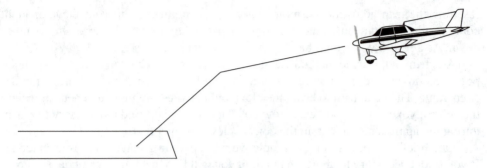

Fig. 6-5. *Obstacle clearance glideslope (power on/off).*

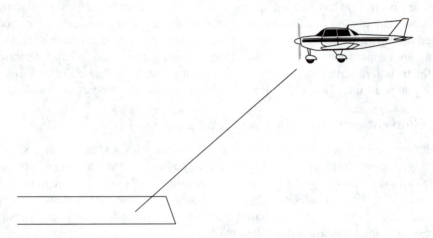

Fig. 6-6. *Obstacle clearance glideslope (constant power).*

might also be tall enough that it is not feasible to set up the second type of approach. Other obstacles in the vicinity of the airport or traffic pattern procedures might also require you to use the first approach method. You will need to determine which approach method works best for you and the plane you are flying.

Depending on the plane you fly, it might be useful to use a sideslip to the runway. I have found that slips can be used to steepen the glideslope if I am going to land long of my intended touchdown point. In the Pitts S2B, I regularly sideslip to help control where I am going to touch down and to keep the runway in view. Planes that are not equipped with flaps, and are certified for slips, often have no alternative but to slip to steepen the approach angle. If the plane has a large wing and lots of lift even with power reduced to idle, it might still be hard to get the correct approach path without slipping the plane. Remember that you will need to transition out of the slip when you get to the correct glideslope. If you need to slip to the flare, you will need to transition out of the slip prior to the flare.

Both approach methods result in a steeper approach to the runway than normal landings. This steeper glideslope also brings along a higher-than-standard rate of descent. This, coupled with the lower approach speed and resulting loss of control effectiveness, means that you will need to plan and execute your flare in the proper manner to avoid "driving" the plane into the ground or flaring too high and dropping the plane in. During the flare itself, you will very likely find that because of the slower airspeed, it will require more elevator travel during the flare. If you begin your flare at the height above the runway you use for a normal landing and use the same amount of control-yoke travel for elevator input, it is possible you will impact the ground at a relatively high rate of descent, possibly bouncing the landing. Begin your flare slightly higher than normal and increase the rate of input for elevator control. When you get the combination down correctly, you will end up in the correct touchdown attitude, whether you are using a wheel landing or a three-point landing. You will need practice to get the sequence down to the point where you can execute the flare and touch down smoothly each time.

Due to the rapid change in pitch during the flare, you might also need to change your rudder control inputs to maintain coordinated flight. Precession and P-factor might cause the plane to yaw during the flare for reasons we have previously discussed. As with the elevator, lower airspeeds will result in lower rudder and aileron effectiveness, so it might take more control input to accomplish the necessary correction.

On the opposite end of the scale from not flaring soon enough, if you flare too high you can stall the plane and drop the landing onto the runway. Because of the slower approach speed, you will need to be careful that you do not overcompensate and add too much elevator input too quickly. This will cause the plane to flare higher than you want it to. With the lower airspeed you use during the approach, there will be less margin of error for recovery from the mistake. If you find that you have flared too high and are slow, add power. Depending on the severity of the error, you might be able to use just enough power to let the plane settle gently to the runway, or you might need to execute a go-around and set the approach up again. With an actual short-field landing, recovering from the high flare might cause you to land so far down the runway that you will not have enough left to stop the plane. Keep this in mind as you recover from the bad flare. There will not be much time to weigh your options. You will need to decide quickly when you are flying the plane near the limits of its landing-distance requirements. If there is any question, though, execute a go-around and set up again.

The previously mentioned strip with the 21-foot-wide runway presented a unique opportunity for landing a Pitts S2B during a recent fly-in there. Planes were landing on the east-west runway, then taxiing to the parking area on the grass just to the south of the runway. The first time I landed there I made my normal approach, slipping down to the altitude at which I transition to a flare. As soon as I did that, the runway completely disappeared and I decided I didn't want to continue that approach because I could not tell where I was in relation to the runway or taxiing aircraft. I executed a go-around and set up for the approach again. I used a steeper slip on final to try to keep the runway in view through the flare. Again, the runway completely disappeared from view. To avoid landing blind with many other airplanes taxiing near the runway, I aborted the landing. On my third approach, at the suggestion of the fly-in sponsors, I landed on the grass just north of the runway without a problem. By keeping the runway in sight to my left and touching down next to it, I had a reference point to work with and I knew I was well away from other aircraft. Never let the pressure to get the plane down the first time distort your judgment. The bad pilot is not the one who executes a go-around and learns from the experience to improve the next approach. The bad pilot is the one who forces the landing in a compromised situation and presents a safety hazard.

There are two basic options for power control on final in short-field landings. Once you can make the runway on final, you can reduce your power to idle. This will result in an increased rate of descent and can be useful in getting rid of unnecessary airspeed as you slow to the correct final-approach value. The other option is to carry power to touchdown, then reduce it to idle. With this option, some of the forward momentum of the plane is provided by the engine's thrust, so as you reduce the power there will be a tendency for the plane to slow down. This can result in a shorter landing distance than if all of the forward momentum of the plane is a result of its gliding speed, such as in the power-off approach we just reviewed. I have not seen any studies that indicate if this method does make any substantial reduction in the runway necessary for landing, but in experiments I did a few years ago in a Cessna 152 it seemed to allow me to get the plane down and stopped in a shorter distance. However, there is a danger that by "hanging it on the prop," if you lose the engine you will not make the runway. This approach method also seemed to work best if the glideslope could be a shallow one, without an obstacle to clear. To have a steep glideslope with power might require you to get into a heavy slip at relatively low altitude. This can be somewhat dangerous if you are at all rusty or unfamiliar with the plane you are flying. In any case, follow the aircraft manufacturer's recommended procedures for short-field landings.

We discussed wheel and three-point landings in chapter 4, and either can be used in short-field landings. As was stated, it depends on your preference, the conditions, and the characteristics of the plane you fly. The flare discussion in previous paragraphs is applicable to both types of landings. However, due to the ability to land at a slower landing speed in the three-point position, you might be able to get stopped in less distance compared to a wheel landing if all other factors are equal. You will need to determine which technique works best for you and the plane you fly.

Once you are on the ground, short-field landings normally require maximum braking. If the plane you fly has flaps, you should retract them immediately after touchdown. This reduces the lift produced by the wings and helps to put more weight on the landing gear, making the brakes more effective. As with flap extension, retraction might cause a pitching moment, so be prepared to compensate for it. Brake application should be firm, but it should not cause skidding of the tires. You might notice a tendency for the plane to want to nose over in heavy braking situations, and this can be true of both wheel and three-point landings. Since the center of gravity is behind the main landing gear, it will attempt to move forward, resulting in a nosing over or turning tendency. Brake application should only be to the point that elevator back pressure can overcome the nose-over tendency.

For example, as you apply brakes you will notice the need to pull back on the control stick to keep the tail in the correct position. As you add more brake pressure, you will need more elevator back pressure. At some point it is possible that elevators will not be able to overcome the nose-over tendency. This is where accidents happen and pilots end up with the plane on its nose. As the plane slows down, elevator effectiveness will be lost and you might need to reduce braking as less tail-down force is available from the elevators. To some degree, wheel landings are more susceptible to nosing over because the center of gravity is higher as compared to the three-point position. As you apply back pressure, you must balance between preventing the plane from nosing over and not forcing the tail to the runway.

I recently flew to an aerobatic competition in the Pitts S2B. There was a strong crosswind from the left, and the runway was just slightly wider than the wingspan of the plane and somewhat short. Not only was I dealing with a stiff crosswind, but also had to contend with getting down and stopped on this small runway. After I touched down, the plane exhibited a strong desire to head to the right, off the runway, so I was trying to keep it somewhat near the center of this very narrow strip and slow it down. I had to keep reminding myself not to overbrake the plane, and I needed to add back pressure on the elevator as I braked heavily to counteract the nose-over tendency. This was the strongest crosswind and the smallest runway I had ever landed this little plane in, and there was no margin for error. During the course of the competition that weekend, I got lots of short-field, crosswind landing practice. This experience showed me that if you stay in control of the plane and are aware of what the plane is telling you, short-field landings in taildraggers can be accomplished without incident.

Short-field landings in conventional-gear planes are fun and challenging. As long as they are performed within the limitations of the plane and pilot, they should not be viewed any differently than a standard approach and landing. In the next section we will cover the short-field flare and rollout for tricycle-gear planes.

Tricycle-gear flare/landing rollout

Short-field landings in tricycle-gear planes are in many ways similar to conventional-gear planes. The purpose is to get the plane down and stopped in as little runway as possible, and many of the techniques are the same. We will start with the final approach and flare, then move on to the landing rollout.

Final approach in a tricycle-gear plane is flown in the same manner as was described for conventional-gear planes. Figures 6-5 and 6-6 both hold as true for tricycle-gear planes as they do for conventional-gear planes. You should make the approach at the speed recommended by the plane's manufacturer and carefully judge your glideslope relative to the intended touchdown point. Short-field approaches must be flown accurately to land in the shortest distance the plane is capable of. If your approach is at too high an airspeed, you will need more runway to get stopped in.

The short-field rate of descent will be higher for tricycle-gear planes than for a standard approach. You will probably need to begin the flare slightly higher than normal. Again, with lower approach speeds it might be necessary to increase the amount of elevator input during the flare to arrive at the correct touchdown attitude at the right altitude. The flare should put you in a slightly nosewheel-high attitude as the main landing gear touches the runway. It can be difficult to get the timing of the flare down just right, and you might find that the plane has a tendency to impact the ground with a higher rate of sink. I have also noticed that many students have a tendency to touch down in a flat attitude, with all three gear hitting the surface at the same time.

You should touch down in a slightly tail-low attitude, with the nosewheel off the runway. Firmly but gently lower the nosewheel to the runway and avoid slamming it to the surface. Usually you want to have the nosewheel on the ground before you begin heavy braking, except in the case of a soft-field landing. You do not want to damage the nosewheel by forcing it to the surface too hard. We will review soft-field landings in the next chapter and discuss why you want to keep the nosewheel off the surface. During short-field landings, pilots make two common errors with regard to the nosewheel. The first is by pushing forward on the control yoke very quickly, which causes the nosewheel to drop rapidly to the runway. The second is to have the nosewheel off the ground as they apply very heavy braking. If the braking is not compensated for with elevator back pressure, assuming there is sufficient speed for control, the plane has a tendency to pivot on the main landing gear, and the nosewheel can strike the ground heavily. By adding enough elevator back pressure to compensate for this, the nosewheel can be held off or lowered at a controlled rate to the runway. If you do not have sufficient airspeed, you will not have the elevator authority to overcome this tendency. Practice short-field landings and try holding the nosewheel off with progressively stronger braking to find out what works well for the plane you fly.

For standard short-field landings, before heavy braking begins, retract the flaps to transfer more of the plane's weight to the main landing gear, and then brake heavily. This helps to reduce the lift the wings are producing and promotes better braking action. As with conventional-gear planes, you want to avoid locking the tires up and skidding them. You will also want to increase the elevator back pressure to avoid the nose-down tendency the plane will experience as the brakes are applied. This helps keep excess weight off the nosewheel. If there is enough momentum and too much braking is applied, the plane might flip itself over onto its back. Be aware of how the plane feels during braking, and don't apply the brakes too quickly or heavily. You should be in complete control of the plane at all times during short-field landings. If you find that the plane is not doing what you want or expect it to during the approach and landing, then your technique might be rusty or you might not be aware of what the plane is telling you. In any case, if you feel out of practice, find a qualified flight instructor to work with you and regain your proficiency.

Some common mistakes made during both conventional and tricycle-gear, short-field landings include not controlling the airspeed during the approach and letting it get too low or too high; not starting the flare at the correct altitude; not retracting the flaps, if any, immediately after touchdown; and not using elevator back pressure during braking. I have also seen some pilots forget to reduce the power to idle, and they touch down with the engine still producing thrust.

SUMMARY

In this chapter we have reviewed a number of different factors related to short-field takeoffs and landings in both conventional and tricycle-gear planes. The key ingredient in successful short-field landings is control. Control the airspeed, rate of descent, flare, and braking to get down and stopped safely in the shortest possible distance. Short-field takeoffs are in no less need of pilot control and awareness. In both flight regimes, you will be flying the plane near its limits of operation, and poor or sloppy technique, lack of knowledge of the manufacturer's recommended operating limitations, or poor judgment can result in a controllable situation getting out of control.

A recent accident at a nearby airport drives home the importance of these points. The pilot was making a southerly approach to a north-south runway that has a bluff at its north end. The pilot made the approach over the bluff at too high an airspeed and too flat an approach. As a result, the plane touched down at the far end of the runway, ran off the end of the runway, and severely damaged the plane. If the pilot had set up a proper obstacle clearance glideslope and maintained the correct airspeed, it is very likely the entire situation could have been avoided. When you are making short-field landings over obstacles, keep the wind conditions in mind. If the winds are light, or the runway is of sufficient length, it might be safer to make a downwind landing as opposed to attempting to clear an obstacle at the other end of the runway. We will discuss downwind landings in chapter 11, but this might have worked out as a better approach in the unfortunate example just described.

Always remember that you have the option to go around and set up the approach again. Don't let the pressure to get it down on the first pass override your judgment. I have seen students become so intent on touching down that they are willing to drive the plane into the runway or land with insufficient distance left to stop.

When I was a low-time taildragger student pilot, I flew a long cross-country with three stops. The first destination turned out to have a strong crosswind almost 90° to the runway. I made four attempts to land the plane, but each time the wind was stronger than I and the plane could deal with. I ended up aborting the landing at that field and flew on to my second stop on the cross-country. It was humbling to admit, but it was the safest thing to do. My flight instructor and the FBO agreed. During any landing, if you do not feel in control of the situation, execute a go-around. As previously stated, only the bad pilot forces a questionable landing situation to "save face." The good pilot reassesses the attempted landing and learns from it for the next time.

SHORT-FIELD TAKEOFFS AND LANDINGS

As you practice short-field landings, understand the correct procedures and be aware of the plane. Different runways, weather conditions, aircraft gross weights, and a host of other factors can change how much runway the plane will need to take off or land. Remember to use the takeoff and landing performance charts to be certain of the runway-length requirements for the conditions you are flying in.

7
Soft-field takeoffs and landings

IN THIS CHAPTER WE WILL DISCUSS SOFT-FIELD TAKEOFF AND LANDING principles and techniques. Soft-field operations are useful for those situations when you are landing on grass, snow, or other soft runway surfaces. They are similar to short-field operations in that you are flying the airplane near the edges of its flight envelope. When operating in these flight regimes, you need to be aware of the plane's limitations and the proper techniques to use. You must also understand the effects that different runway conditions will have on your takeoff and landing. Many of the concepts and techniques for short-field landings that were reviewed in chapter 6 will apply directly to topics we will discuss in this chapter.

As always, we will cover soft-field takeoffs and landings for both conventional- and tricycle-gear aircraft. Some of the special hazards associated with soft-field work will also be reviewed. When you have completed this chapter, you will have a solid understanding of the principles behind soft-field takeoffs and landings. Keep in mind that these are principles and are not able to cover all situations. You should use the manufacturer's recommended techniques in addition to what your own experience, fellow pilots, and flight instructors have taught you. Let's begin with a review of a previously discussed topic that is directly related to soft-field takeoffs—ground effect.

GROUND EFFECT/STALL

In chapter 4 we reviewed the factors related to ground effect. Ground effect heavily influences soft-field operations, in particular soft-field takeoffs. Understanding ground effect, and how you can use it to your benefit, can help you avoid the hazard associated with its misuse—the stall. As the plane comes within a wingspan's distance from the runway, the airflow around the plane is changed by the ground's interference with it, resulting in a reduction of induced drag. Induced drag drops dramatically as the plane descends closer to the surface. At one-tenth of the wingspan's distance, induced drag is reduced by 47.6%. A number of beneficial effects result from the lowered drag. The wing, now in ground effect, requires a lower angle of attack to achieve the same coefficient of lift as the same wing outside of ground effect. Less thrust is also necessary to maintain a given airspeed as compared to the plane not in ground effect. Local air pressure becomes higher near the static source, resulting in a lower indicated airspeed near the runway's surface. During landing, the effects can cause the plane to float as airspeed and excess lift are bled off prior to touchdown (*Flight Training Handbook*, p. 271).

Ground effect also affects takeoff. If ground effect is not understood, this can lead the unwary pilot into believing the plane is ready to take off and climb when, in fact, it is not. Under these circumstances, the plane will often settle back to the runway, fly just above the ground and be unable to climb, or stall as the pilot attempts to force the plane to climb out of ground effect when it is unable to. During takeoff, the increase in static source pressure, which results in a reduced indicated airspeed, might make it seem as though the plane is able to fly at airspeeds below its normal range. As the plane leaves ground effect, the indicated airspeed will increase due to the change in static-source pressure. The induced drag generated by the wing will also increase. Increased drag will require additional thrust to maintain a given airspeed, and increased drag will require an increased angle of attack to produce the same coefficient of lift that it did while in ground effect. There will also be a nose-up pitching moment and reduced aircraft stability.

Under normal takeoff airspeeds, the results of leaving ground effect after lift-off will be minimal. Your plane will have sufficient airspeed to continue to climb, and there might be no noticeable changes in aircraft performance. During takeoffs that are at airspeeds much closer to the aircraft's stalling speed, you will find that ground effect plays a major role during the initial moments after the plane leaves the ground. If you force the plane off before it is ready to fly, ground effect might allow the plane to get airborne, but as you try to continue the climb you might find yourself "behind the power curve." Each time you attempt to climb out of ground effect, the increased induced drag and the need for additional lift and thrust can cause the plane to settle back toward the runway. At this point you do not have adequate engine thrust or lift to continue to climb. Under the right conditions, you can allow the plane to accelerate in ground effect, then climb away normally. There have been numerous accidents, though, where the pilot of a heavily loaded plane, or one flying from a high-density altitude, has managed to get the plane off the runway but has been unable to gain additional altitude. In many cases pilots continue to fly along in ground effect, well past the end of the runway, until they impact the ground or an obstacle in their flight path. There are also many documented accidents where the pilot attempted to force the plane to climb out of ground effect, resulting in a stall with insufficient altitude to recover.

Ground effect can be used to your advantage, though, so let's discuss how it plays a beneficial role in soft-field takeoffs. Soft-field takeoffs take advantage of the positive benefits of ground effect to allow the airplane to become airborne at a slower airspeed. Lower induced drag and lower thrust and lift requirements let you get the airplane off the ground at relatively slow airspeeds and short runway distances. Becoming airborne as quickly as possible during soft-field takeoffs reduces the negative impacts of drag, slower acceleration, and longer takeoff distances, which are generated as a result of the soft-field runway's surface.

Soft-field runways can come in many varieties, but they are often grass or dirt runways. They can also be hard-surfaced runways covered with snow. Any runway surface that generates a large amount of drag on the plane's tires, inhibiting its ability to accelerate at a normal rate and increasing the takeoff distance, can qualify as a candidate for a soft-field takeoff. In some cases, tall grass, soft ground, or deep snow can prevent an airplane from reaching takeoff speeds using normal takeoff techniques, and the pilot runs off the end of the runway waiting to reach takeoff speed. Using soft-field takeoff techniques, you get as much weight off the main landing gear as soon as possible, allowing the plane to reduce the drag generated by the soft field. The key to successful soft-field takeoffs is an understanding of ground effect, the performance capabilities of the plane, and proper soft-field takeoff techniques.

TAKEOFF

In this section we will discuss soft-field takeoff techniques. We will begin with flap and trim settings, then move on to takeoff techniques for conventional- and tricycle-gear aircraft. Please remember that each plane will be different in its soft-field takeoff characteristics. As you practice soft-field takeoffs, you should be aware of the aircraft and how it "feels" during the takeoff. Control inputs that work well for one model aircraft might need to be modified for another. In fact, the same plane might require slightly different control inputs based on aircraft weight, power settings, weather conditions, and many other factors. If you feel at all out of practice, or would like some additional clarification related to soft-field takeoffs, find a qualified flight instructor to work with you and help improve your performance.

Flap settings

There are a number of factors to consider in the use of flaps during soft-field takeoffs. First among them is the manufacturer's recommended flaps setting during the takeoff.

Like short-field takeoffs, the recommended flap settings might vary from the use of no flaps during the takeoff to one or two notches. If recommended for the plane you fly, the use of flaps will help generate additional lift, allowing the plane to become airborne at a slower airspeed. Do not exceed the recommended settings, though. Any additional flaps above the recommended settings of the manufacturer might result in additional drag and a potentially longer takeoff run.

The second factor to consider is the runway surface. If you are flying from a strip that has mud, large puddles, snow, or any debris that can be thrown up by the landing-gear wheels, you should be cautious in the use of flaps. There have been many documented cases where, during a soft-field takeoff, objects were thrown from the wheels into the extended flaps, causing damage. In cold weather, snow or water can be thrown up onto the flaps' tracks, which are exposed by the extended flaps. The water can freeze there and prevent the flaps from being retracted or extended. As you look over the soft-field runway prior to your takeoff, determine if the conditions are such that it might be possible for this to happen. Low-wing planes are even more susceptible to debris damage than high-wing airplanes, since the flaps and flaps' tracks are much closer to the main landing gear.

In the event that you elect not to take off with extended flaps for this or other reasons, you will need to be aware of how that affects the takeoff performance of the plane. It is possible that you will need to accelerate to a higher airspeed before the plane will become airborne, resulting in the need for additional runway. In a soft-field situation the additional drag might result in a substantial increase in the takeoff distance. If you are heavily loaded, or if the runway surface generates a great deal of drag, you might find that you cannot lift off without the recommended flaps settings. In this situation you might not have a choice in the use of flaps, other than to not take off until conditions improve.

Trim settings

As with other types of takeoffs, you should use the recommended setting from the aircraft's manufacturer. In most cases this will be a neutral trim setting, which is used for most takeoff scenarios. You should not add additional nose-up trim in anticipation of the need for additional elevator back pressure during the takeoff run. You might be able to do this and not have problems, but there are situations where excessive nose-up trim could result in problems. It is possible this could lead to the nose of the plane coming off the runway before sufficient airspeed has been achieved or for the nose to rise too high during the climb out after lift-off. If your center of gravity is located toward the rear of the "envelope," it is possible you might not be able to overcome the nose-up tendency resulting from the combination of nose-up trim and aft CG.

Excessive nose-down trim can generate some unique problems of its own during a soft-field takeoff, most notably with tricycle-gear aircraft. As you hold back pressure during the takeoff run to lighten the weight and drag on the main landing gear, nose-down trim will attempt to offset your back pressure inputs and force the nose down. In both conventional- and tricycle-gear aircraft, the additional drag due to the extra weight on the landing gear will reduce the rate of acceleration and increase the takeoff distance. The additional drag might also generate a greater nose-over tendency. In conventional-gear planes, this could result in the tail of the plane rising and the propeller striking the runway. In tricycle-gear planes, it could cause damage to the propeller and nosewheel assembly. Improper techniques, such as incorrect use of trim, open up the possibility for avoidable problems to arise during the takeoff.

Conventional-gear ground roll and rotation

In this section we will cover the proper techniques for soft-field takeoffs in conventional-gear planes. The purpose of soft-field takeoffs is to quickly remove as much of the aircraft's weight from the landing gear as is possible, reducing the drag generated by a soft runway surface such as mud or snow. By taking advantage of ground effect, the plane can lift off and fly in ground effect, continuing to accelerate to a sufficient airspeed to allow it to climb. This type of lift-off allows the plane to escape the drag generated by the runway and reduces the potential for objects from the runway to be thrown up against the flaps and wings. Soft-field takeoffs also present a stronger-than-normal tendency for a taildragger to nose over. The sooner the plane is free of the runway, the less the chance that this can take place.

In this section we will review the standard techniques you should use, in addition to the potential problems you should be aware of during the takeoff run. These are general techniques that work well for most conventional-gear aircraft. However, if your plane's operations manual specifies a different set of procedures, then use those. It cannot be overemphasized that if you are new to soft-field takeoffs, or have not performed them for a long period of time, find a qualified flight instructor to ride with you until you become proficient. Let's begin by covering the rationale for soft-field takeoffs and what purposes they should be used for.

Soft-field takeoffs can be from a variety of runway surfaces. A common trait among these runways is that they produce a large amount of drag on the wheels and landing gear, which slows the rate of acceleration of the plane during the takeoff roll. Some examples of soft-field takeoff surfaces I have flown from include tall grass runways, soft grass runways, and snow-covered runways. As previously stated, soft-field takeoff techniques and ground effect are used to your advantage to quickly get as much weight off the wheels as possible. When properly executed, you will get the wheels off the runway's surface, fly in ground effect until you accelerate to V_X or V_Y, then climb. By flying a few feet above the ground, you can get away from the effects of the runway's drag, even in moderately tall grass.

Before attempting a soft-field takeoff, you should visually inspect the runway by walking along it. Look for any objects, hidden potholes, soft spots, or other things that present a danger to the aircraft. Due to the drag they generate, potholes, muddy areas, and deep snow (to name a few items) all pose the potential for causing the plane to nose over, and they also have a tendency to pull the plane off its track down the runway. It is much better to be aware of these before the takeoff begins than to find out during the takeoff roll. In many cases you will be able to steer the plane around these potential hazards if you know they are there ahead of time.

After you start the plane, you will want to do the runup in the usual manner on the most solid surface available. After you complete the runup and before you take the runway, verify that there is no traffic landing or taking off. As you taxi onto the runway, keep your power up and do not slow to a stop as you align yourself with the runway, as you would in normal takeoffs. If the runway's surface is sufficiently soft, it is possible for the plane to become mired and not be able to move forward or accelerate. As you taxi, you should also keep the control stick back sufficiently to keep the tail down and help prevent any nose-over tendency. Figure 7-1 illustrates the elevators in the correct position for the beginning of the takeoff run.

Fig. 7-1. *Soft-field takeoff elevators (conventional gear).*

Many taildraggers present a special challenge for soft-field takeoffs due to the limited forward visibility they have over the nose. As you turn onto the runway, make sure you are parallel with its centerline. If you are not, it is quite possible to end up heading off the runway during the takeoff roll. As you complete the turn onto the runway, bring in full power and keep the control stick back enough to prevent the plane from nosing over. During the takeoff run, keep the stick aft until the plane begins to become light on the landing gear, then gently ease off some of the back pressure. As the plane gains speed, it might be necessary to release some back pressure on the stick to prevent the plane from lifting off at too low an airspeed. The plane should be lifted off in a roughly three-point attitude, then leveled off just above the runway's surface to allow it to accelerate in ground effect. If you have too much aft control stick during the takeoff run, the plane might climb too steeply and stall. To be executed properly on a consistent basis, soft-field takeoffs require a "feel" for the plane and a finesse on the controls. After the plane has accelerated to the correct airspeed, you can establish a climb and continue to fly away.

The aft stick during the takeoff run accomplishes two objectives. The first is to overcome any nose-over tendency the plane might exhibit due to the soft surface. The second objective is to be in the three-point attitude as the plane accelerates. This increases the angle of attack of the wings and generates greater lift at a lower airspeed. As previously discussed, as the plane becomes light on the landing gear, ease the stick progressively forward to allow the plane to lift off in a three-point attitude, then level off just above the runway surface.

During the takeoff run, there might be a tendency for the plane to be pulled in one direction or another by soft spots on the runway. This requires you to compensate with steering and brakes to maintain the proper heading. A weekly aviation television show has displayed footage of P-51 Mustangs taking off and landing on a tremendously wet, muddy runway with sheets of water splashing up from the landing gear over the leading edges of the wings. The planes bounce and splash along until they become airborne, demonstrating just how bad a soft-field takeoff can be. While I don't recommend that anyone attempt takeoffs in those types of conditions, that film clip demonstrates that takeoffs can be accomplished in much less than ideal conditions if the proper techniques are used.

You should be aware of a number of situations and problems that can be encountered during the takeoff. There might be a need for a great deal of right rudder during the takeoff. Due to the slow airspeed and high angle of attack at lift-off, some planes will exhibit a very strong tendency to turn left during the takeoff roll and lift-off. Don't let the plane get away from you and start off to the left. Immediately compensate for any turning tendency with right rudder. Remember to compensate with ailerons after lift-off to keep the wings level as you add rudder. You should also be aware of how much runway you have remaining during the takeoff run. Takeoff distance charts can only give you a very rough idea of the runway requirements for a soft-field takeoff. Hard-packed snow or other relatively hard surfaces will produce much less drag than a very soft field, deep snow, or tall grass. The latter examples might require a great deal more runway due to the reduced acceleration they cause. As you roll down the runway, be aware of your rate of acceleration, airspeed, feel of the plane, and runway remaining. You will need to monitor the runway remaining and whether you are going to be able to lift off before you are out of runway. Each situation can be different, and the better you know the plane you are flying, the more adept you will be at making the go/no-go call during the takeoff run. If you must abort the takeoff, reduce power to idle, retract flaps if necessary, keep the control stick back to help keep the tail on the ground, and brake as the situation warrants. You might find the plane slows very quickly due to the drag of the runway. In this situation do not overbrake and cause the plane to nose over into the runway.

Some common errors made during a soft-field takeoff include stopping on the runway after turning onto it (which can cause the plane to become stuck), not keeping the control stick back during the takeoff roll to keep the tail down, not correcting with right rudder, not maintaining the runway heading, and not leveling off after lift-off to allow the plane to accelerate in ground effect. The lift-off itself requires a deft touch on the controls, especially the elevator, to keep the plane from climbing out of ground effect at too steep an angle. Do not overcompensate with forward stick and settle back to the runway's surface. Not having enough back pressure might cause the plane to stay on the runway longer, increasing the length of the takeoff run. If you are flying in windy conditions, it might be a good idea to maintain a higher airspeed before lift-off to give yourself an airspeed buffer to provide more control compensation. In this situation you should use less back pressure as the airplane becomes light on the landing gear. This allows the plane to stay on the runway until it accelerates to a higher airspeed.

If you fly a taildragger, make sure you practice soft-field takeoffs in it. Consistently performing them in a proficient manner demonstrates your finesse, understanding, and control of the airplane. In the next section we will discuss soft-field takeoff techniques for tricycle-gear planes and contrast those with conventional-gear techniques.

Tricycle-gear ground roll and rotation

Many of the principles we covered for conventional-gear, soft-field takeoffs hold true for tricycle-gear planes as well. The technique differs slightly because now there is a nosewheel instead of a tailwheel, which results in somewhat different aircraft behavior during the takeoff. In tricycle-gear planes, the correct technique results in the plane lifting the nosewheel from the runway surface as early as possible. This helps transfer the weight from the main landing gear more quickly by increasing the angle of attack, and thereby, lift. This also reduces the drag that is generated by the nosewheel while it is in contact with the soft runway surface. The lower landing gear drag lets the plane accelerate more quickly, and keeping the nosewheel off the runway reduces any nosing-over tendency that might be caused if the plane becomes mired during the takeoff run.

For the reasons we previously discussed, it's a good idea to inspect the runway before attempting the takeoff. After the runup, do not slow down as you taxi onto the runway. Keep some power in with the control yoke pulled fully aft as you taxi to the runway and turn to align yourself with the runway centerline. Be sure that you are parallel with the centerline as you bring in full power. Figure 7-2 shows the elevator with the control yoke in the full aft position as the pilot applies full power. Most tricycle-gear planes have better forward visibility than conventional-gear planes, which makes the task of maintaining the correct track during the initial takeoff roll somewhat easier. As the plane accelerates and the elevators gain effectiveness, the plane will have a tendency to pitch the nose up. At this point you begin to balance the plane on its main landing gear, holding the nosegear off the runway surface, yet keeping it low enough that the plane does not leave the ground before it is ready. This is done using the ele-

Fig. 7-2. *Soft-field takeoff elevators (tricycle gear).*

vators and reducing back pressure as they become more effective. Inputs to the elevator should be smooth and small, with the nose maintaining a relatively constant distance from the runway's surface and the plane a relatively constant angle of attack. Having the nose up might reduce your ability to see the runway ahead and make it more difficult to track straight down the runway. Without the nosewheel on the ground, nosewheel steering will not be available and you will need to use brakes and rudder to hold your heading.

As the plane becomes lighter on the main landing gear, you will be close to flying speed in ground effect. When the plane leaves the ground you should reduce elevator back pressure just enough to level off above the runway surface, allowing the plane to continue to accelerate. Then climb away at V_X or V_Y, depending on the situation. If you need to clear an obstacle, climb out at V_X. Otherwise, use V_Y or the recommended climb speed for your aircraft. As the plane becomes lighter on the landing gear, you will need to adjust your inputs to ease the plane off the runway in a controlled manner. If you have too much elevator back pressure, the plane might climb out steeply and stall or settle back to the runway as it leaves ground effect. If you have insufficient elevator back pressure, the plane will require more runway for the takeoff as a result of the lower angle of attack and resulting loss of lift from the wings. As you can see, experience and practice become an important part of successful, proficient soft-field takeoffs. You must develop a feel for the plane you are flying and adjust your technique to a given set of conditions or different airplanes.

Some tricycle-gear planes that I have performed soft-field takeoffs in require a relatively large amount of right rudder during the takeoff roll due to the high angle of attack and slow airspeed. This is very noticeable as the nosewheel comes off the ground. Be prepared for this. At a local airport in recent years, there have been two injury-free accidents caused by lack of necessary right rudder application in the nose-high takeoff roll attitude. In one case, the plane, flying from a grass strip, became airborne as it swerved left off the runway and clipped a small tree. The aircraft was severely damaged as it came to rest again. In the other case the plane was using a hard-surface runway for soft-field takeoff practice. Due to lack of adequate right-rudder application, the aircraft swerved left as the nose was raised. The left main gear encountered soft, muddy ground as it left the runway and, to make a long story short, the plane ended up on its back next to the runway. Both of these situations could have been avoided if the proper techniques had been used and the reaction of the planes during soft-field takeoffs had been anticipated.

Soft spots, ruts in the runway, potholes, and other hazards will try to drag one or both main landing gears during the takeoff run. Through the use of brakes and rudder, you will need to maintain directional control as the plane is accelerating. With the nosewheel off the runway, there will also be a tendency for it to drop back to the surface when you encounter these hazards. This might require that you use additional back pressure on the control yoke briefly to keep the nose at the proper angle. As previously stated, once the nose is off the ground, you might find it more difficult to see the runway ahead of you. At this point you might need to use the sides of the runway, much like taildragger pilots, to maintain runway orientation. This nose-high attitude might also block your view of any potential hazards in front of you, so the runway inspection gives you an idea of where any might be.

You might find it necessary to take off on one side of the runway, as opposed to using the middle during the takeoff run. If the runway has become rutted from other aircraft use, has low spots that are flooded or very soft, or has any other hazards that do not span its entire width, it might be safer to keep the takeoff roll away from that area. I have used grass strips that had holes dug in them by our furry, little, woodland friends. To avoid these potential tire-grabbing hazards, I stayed on the side of the runway away from them during my takeoff runs. If you do this, be careful that you are not setting yourself up for other hazards, though. For instance, if there are trees, corn fields, buildings, poles, or any other objects the plane might strike next to the runway, you do not want to get so far over that a wing can hit them. In any situation you must use your best judgment as you weigh the best course of action. The worst thing you can do is become complacent in your approach to the takeoffs.

There are several common mistakes made during the soft-field takeoff in tricycle-gear planes. Not keeping the plane rolling as you turn onto the runway and letting the aircraft stop can cause the plane to become stuck or increase the takeoff roll a considerable distance. Maintaining too much or too little elevator back pressure is also a common mistake until you develop a feel for the plane you are flying. Insufficient rudder to overcome the left-turning tendency can result in the plane veering during the takeoff roll or leaving the runway in a worst-case situation. Finally, not leveling off in ground effect, but trying to continue to climb after lift-off, can result in the plane settling back to the ground. One final note: at gross weight, or days with high-density altitude, it might not be possible to get the airplane out of ground effect after you lift off at minimum airspeed. If necessary, let the airplane accelerate to a higher speed before you take off. If the conditions warrant, you might have to reduce the weight of the plane or wait for more favorable conditions.

LANDINGS

Soft-field landings are fun to execute. The intent is to touch down at minimum airspeed as gently as possible. This reduces the potential for the gear digging in to the runway's surface and nosing the plane over. Depending on the surface you are landing on, there might be a strong tendency for the plane to nose over. When I worked as a line boy many years ago, I saw a Cessna taildragger land on a sod strip in the early spring. The runway had not been opened yet, and it was very soft. As the plane touched down, the tail came up and the prop started cutting through the soft ground. The pilot got the tail back down before the plane nosed over completely, and the pilot prevented serious damage to the plane. After the excitement died down and the plane was towed to the hanger for inspection, I went out to the sod runway and examined the cuts in the ground where the propeller had cut through it. The slices were 1 or 2 inches deep, evenly spaced along the ground. The pilot was lucky that the runway had been so soft. The prop was not damaged and no engine damage was found, either. But this incident demonstrated to me that it is quite possible for soft-field landings to get out of control if the pilot does not anticipate and prepare for them.

If you make soft-field landings, always be prepared for the nose-over tendency that is inherent in them. This is even more true for landings than for takeoffs. You are touching down at a relatively high speed, and the drag from the runway's surface can have an immediate braking tendency and cause the plane to pitch forward. The faster your airspeed, the stronger this tendency. The speed gradually increases during soft-field takeoffs, as does the nose-over tendency. But the deceleration is very abrupt during soft-field landings, and this requires anticipation on your part to prevent any accidents. In this section we will review soft-field landing techniques for conventional- and tricycle-gear airplanes. Each has somewhat different results in the plane's touchdown, but if properly executed, these techniques can reduce the potential for nosing the plane over. One more caution, though: these are generalized techniques that have worked well for the aircraft I have flown over the years. If the manufacturer of the plane you fly recommends a different set of procedures, you should use them. You should also have a qualified flight instructor ride along with you and critique your soft-field landings until you gain proficiency in their execution. In the following sections we will review the flaps settings, trim settings, and soft-field landing procedures. Some of the concepts have been covered previously and will be mentioned only briefly to help refresh your memory.

Flap settings

Using flaps in soft-field landings helps to reduce the stall speed of the aircraft and, therefore, the touchdown speed. Be aware, though, that lowered flaps present a target for debris (such as mud, snow, or ice) that might be thrown up by your landing gear. If you are not familiar with a runway that you are executing a soft-field landing into, it is a good idea to do a low flyby down the runway, conditions permitting. Observe the runway surface and look for any potential hazards. You should be looking for areas where water has gathered, muddy areas, low spots, or other areas that might cause problems during the landing and rollout. The flap settings recommended by your aircraft's manufacturer for soft-field landings should be used whenever possible. If you find that the field you are landing at presents a hazard for debris being thrown up into the flaps, you might try landing with the flaps in the recommended position and retract them immediately after touchdown. It is also possible that partial flaps can be used in place of full flaps to present a smaller target for debris to strike.

If you decide to use flaps, follow the normal flap-extension procedures covered previously in this book. As you extend the flaps, be aware of the effect they have on the glideslope and airspeed. Control of both of these factors is as important in soft-field landings as it is in short-field landings. Fluctuating airspeeds and glideslopes make it more difficult to hit your touchdown point. You might find it necessary to lower the nose or increase power to maintain the correct airspeeds. Since you are making the approach at minimal airspeeds, there will be little room for airspeed loss, and it is necessary for you to anticipate the reaction of the plane for each flap increment extension.

In the event that you are unable to use flaps during the landing, adjust your approach speed upwards accordingly. This will also result in a higher speed at touchdown and possibly stronger drag on the landing gear and stronger nose-over tendencies. Be prepared for this, and compensate with additional back pressure on the control stick. Use the elevators to help keep the tail down. The higher touchdown speed might provide additional airflow over the elevators and make more tail-down force available.

Trim setting

Trim should be set in the normal manner described several times previously in this book. Use trim to help you "lock" the correct airspeeds during the approach by trimming for the speed you want to maintain. When it is correctly set, the plane should maintain a nose-pitch angle that holds the necessary airspeed without a great deal of elevator-control input pressure from you. Depending on the weather conditions and center of gravity of the plane, trim settings can vary to achieve the airspeed you want to hold. But do not input too much nose-up trim in anticipation of the elevator back pressure necessary during the soft-field touchdown. Elevator trim should be used to reduce your work load during the approach to landing.

Conventional-gear flare and rollout

Soft-field landings in a taildragger are very similar to short-field landings. You will normally want to complete them in a three-point position because this offers the best resistance to the nose-over tendency that soft-field landings present (as was documented in the incident at the start of the chapter). You should also be attempting to touch down at the slowest airspeed that safety allows. In conventional-gear planes, this will again normally result in the plane being in a three-point position because this offers the slowest touchdown speed at, or just above, the stall speed. (See Fig. 7-3.)

Fig. 7-3. *Soft-field landing attitude (conventional gear).*

Final approach speeds will be approximately the same as for short-field landings, which tend to offer the lowest forward speed at touchdown. You should pick your touchdown point on the runway and fly to it using the touchdown-zone targeting techniques described previously in this book. In some taildraggers, you might have trouble maintaining line of sight with the runway. This is because of the tendency of many taildraggers to be blind over the nose. If this is the case, you might need to fly a power-off approach, which normally allows you to lower the nose and provides a less-obstructed forward view. Another method that increases forward visibility is to slip the airplane during final approach. During the approach, set flaps in the normal manner and, in most cases, you will want to use full flaps to help reduce your airspeed at touchdown as much as possible. Due to the lower airspeed on final approach, you will find there is reduced elevator effectiveness, and during the flare it might be necessary to input greater control movement to achieve the desired three-point position. Once in the three-point position, hold the plane off the ground as long as possible to further reduce the airspeed. The touchdown should be as light as possible to reduce the chances of the plane's landing gear digging into the runway more than necessary.

As you approach the runway, ground effect can again be used to your advantage, depending on your plane's flight characteristics. In the same way described for take-offs, as the plane nears the ground the induced drag that is generated is reduced and the plane requires less lift to maintain flight. This effect will vary from one type of plane to another, but having this additional lift available can allow you to further reduce the plane's speed before it touches down. As you gain experience with a plane, you will develop a feel for it as it enters ground effect during the flare.

As you touch down, you should have the control stick in the full aft position, offering the most tail-down force available to overcome any tendency for the plane to nose over. Be prepared for any directional changes caused by the soft surface as one landing gear or the other encounters soft spots on the runway. As much as possible, maintain your runway alignment. In the three-point position, you might not be able to see the runway directly ahead of you, and it might be necessary to use peripheral vision to see the sides of the runway. This can present a challenge, depending on the condition of the runway. Runway conditions might also require you to steer around hazards during the rollout in much same way as previously described for soft-field takeoffs.

In some cases, soft-field landings offer one type of advantage, though. Depending on the length of the runway and the type of surface you are landing on, braking might not be necessary during the rollout. Drag from the runway surface might generate enough braking effect to slow the plane quickly. In fact, after touchdown it might be necessary for you to add power and taxi the plane along the runway without slowing too much or coming to a stop to avoid becoming stuck. If this is the case, be sure to keep the control stick in the full aft position as you taxi.

Soft-field landings require a level of finesse that's similar to the level required for takeoffs, and a good feel for the plane is just as necessary. In a well-executed landing, you will hold the plane off the runway in a three-point position until it stalls and gently touches down. To execute consistently, this requires an awareness of the correct attitude and height above the runway. To achieve this, a proper flare is required to place the aircraft at the correct height and airspeed. If you flare too late, your rate of descent will be too high, as will your airspeed, and the plane will have a greater tendency to "sink" into the soft runway surface, generating additional drag. Flaring too high will result in the plane dropping onto the runway, again sinking in further as a result of the higher rate of descent. As you can see, all of the factors that go into making good, consistent landings apply in this case as well. By now the pattern should be forming. To make landings correctly on a consistent basis, all of the techniques that are part of it must be done well. Mistakes during any phase can be compounded as the approach continues, resulting in a substandard landing.

Many of the errors common to short-field landings apply to the soft-field version. Poor airspeed control during the approach, inconsistent glidepath, incorrect flaring altitude, and touching down too hard are among the common errors. Not keeping the control stick fully aft during the rollout is another frequent mistake. Letting the plane come to a stop on the runway, then getting stuck is also not uncommon. If you find the tail coming up, even with full-back elevator, a brief burst of power from the engine might generate enough airflow over the elevators to help them keep the tail down. Be sure to have the stick fully aft, though, or the nose-over tendency might be increased. Remember that you might need to retract flaps, if equipped, to avoid objects being thrown up into them by the wheels.

Tricycle-gear flare and landing rollout

There are many similarities between tricycle- and conventional-gear soft-field landings. The approach, airspeed considerations, use of flaps, and control inputs are all based on the same principles for both aircraft types. The result is that the approach to landings is very much the same as described in the previous section. The major difference is in the flare and the attitude you want the plane to achieve at touchdown. Your flare should place a tricycle-gear plane in a nose-high position, touching down on the main landing gear at or just above stalling speed. A well-timed flare, combined with proper airspeed, are crucial to consistent, well-executed soft-field landings. If your airspeed is too high, there will be additional stress on the plane's landing gear from the additional drag. If it is too low, your rate of descent will probably be higher than normal, resulting in a harder impact at touchdown and more stress on the landing gear. Flaring at the correct height also reduces stress on the landing gear for the same reasons. If you flare too high, the plane will stall well above the runway and drop in. Flaring too low results in an excessive rate of descent at touchdown. Knowing your airplane's characteristics during the flare and being able to accurately judge your height are very important.

Ground effect plays the same role in tricycle-gear planes as with taildraggers. By using the reduced drag and lower lift requirements to your advantage, you can further reduce the speed you will touch down at. This needs to also be weighed against the length of the runway you are landing on, though. If you are landing on a short runway, you do not want to "float" down it while trying to get rid of excess airspeed. Again, we can see that airspeed control is very important in keeping touchdown speeds as low as possible while using a minimum amount of runway. Like taildraggers, experience flying your plane will help increase your "feel" for changes in flight characteristics as your craft enters ground effect.

At touchdown the nosewheel should be off the runway surface, with the control wheel in the full-aft position. The proper landing attitude is illustrated in Fig. 7-4. The combination of the nose-high touchdown and aft control wheel keeps the nosewheel off the ground and helps to reduce the nose-over tendency the plane might have. As the airspeed falls off, gently lower the nosewheel surface. However, once the nosewheel is down, maintain the control wheel fully aft to keep as much pressure as possible off the nosegear. It might be necessary to use right rudder to keep the plane tracking straight down the runway during rollout to compensate for P-factor, which can result from the

Fig. 7-4. *Soft-field landing attitude (tricycle gear).*

high angle of attack at touchdown. Use rudder and/or brakes to maintain your directional control until the nosegear touchdown. Then use nosewheel steering, if available.

The plane's nose-high attitude might prevent you from maintaining a clear view of the runway ahead. In this case it might be necessary to use peripheral vision to track straight ahead, as we discussed in the conventional-gear section. Do not let the plane come to a rest on the runway. If necessary, add power to keep the plane rolling until you are clear of the runway and reach a surface you can stop or park the plane on.

To avoid unnecessary repetition, only differences between taildragger and tricycle-gear aircraft have been reviewed in this section. Common errors include not keeping the nosegear off the ground after touchdown, not using enough right rudder with the nose up to stay parallel to the runway centerline, and not keeping the control wheel back after the nosewheel touches down.

SUMMARY

In this chapter we have reviewed the subtleties of soft-field takeoffs and landings. Soft-field takeoffs require a smooth touch on the controls and a feel for the airplane you are flying. This allows you to safely get the airplane off the runway and remain in ground effect until you have accelerated sufficiently to fly away normally. Overcontrolling can cause the plane to settle back to the runway or stall if you attempt to climb out of ground effect too quickly. By practicing soft-field takeoffs, you can perfect these techniques and learn to execute soft-field takeoffs and landings consistently and proficiently. You might also want to consider practicing flight at minimum controllable airspeeds at a safe altitude to help you increase your feel for the plane's flight qualities at slow airspeeds. As we have already discussed, the controls will be less effective and will require additional input to achieve the desired results.

In many ways, soft-field landings are very similar to short-field landings. Many of the same principles, such as airspeed control, touchdown-point judgment, and glide-slope control, apply to both. There are also many similarities between the soft-field landing techniques for conventional- and tricycle-gear planes. The major difference is that you want all three wheels on the ground at touchdown for taildraggers, and you want the nosewheel off the ground for tricycle-gear planes. The end result for both, though, is to land at stall speed and prevent the plane from nosing over. As you practice soft-field landings in either type of aircraft, keep in mind that you want to touch down at the slowest possible airspeed and as gently as possible to reduce the tendency for the plane's gear to dig into the soft runway surface. Also remember to keep the elevator control in the full aft position to help keep the tail down and prevent the plane from nosing over.

The type of soft field you are landing on should be considered as you make your takeoff and landing. Many times soft-field runways can be grass-covered, rolling ground, or on a hill. In these situations, low areas can present a greater hazard as a result of water, snow, or mud gathering there. When you are landing in these conditions, you should try to plan your touchdown in higher areas and avoid the low spots whenever possible. As you can see, picking and hitting your touchdown point becomes very important in these situations. When the runway is on a hill, make your landing away from the softer areas that might be located at the lower end of the runway. In some cases, downwind landings might work better at staying away from softer, more hazardous areas, so keep that in mind as an option. When landing on hilly, rolling terrain, you will need to plan your flare with the terrain you are touching down on. If the ground is uphill at the point of touchdown, the ground will be coming up to meet the plane. In this scenario, you should be trying to work parallel to the ground, rather than flying along level and having a harder impact. If the ground is downhill, again, work parallel to it. If you flare and keep level flight while the runway drops away from you, the result might be a drop of several feet to the now-distant runway below you. The main point to remember is to stay flexible in your takeoff and landing procedures, and modify them to best meet a given situation.

As you practice soft-field takeoffs and landings, start with normal runways and perfect your technique before you move on to practice using an actual soft runway. However, once you can consistently and accurately perform them, you should move to progressively softer runways, when available, to further improve your abilities. Be careful, though, and know what the limitations for you and your airplane are. Under some conditions it will not be possible to land or take off safely. Always use your judgment and keep safety as the number one criteria when making a decision to take off or land. As will be explained in chapter 11, soft-field landing techniques can also play an important part in successful forced landings.

8
Crosswind takeoffs and landings

IN THIS CHAPTER WE ARE GOING TO REVIEW CROSSWIND TAKEOFF AND landing principles and techniques. I enjoy crosswind work and feel that it demonstrates how well a pilot understands handling the airplane. This becomes very true when the crosswind is coupled with other types of procedures, such as short- or soft-field work. When you are able to still land on the target touchdown point, on the centerline, and maintain a track down the runway in a stiff crosswind, you gain a great deal of confidence in yourself and the airplane.

We will again cover techniques for conventional- and tricycle-gear planes, in addition to the other standard topics. We will also discuss crosswind taxi techniques which, when properly performed, help you maintain control of the plane as you taxi in strong winds. As you review this chapter, keep in mind that you should never exceed the demonstrated crosswind component for your plane. Reduce the maximum crosswind you operate in to even less if you are new to a plane or have not performed crosswind takeoffs or landings in some time. Practice is the key to being proficient, and the more crosswind work you get, the better you will become at it. Finally, although it has been frequently stated, if you are out of practice or new to a plane, find a qualified flight instructor to work with you to improve your crosswind operating techniques.

CROSSWIND TAXIING

One of the most ignored areas of crosswind flying is crosswind taxiing in a plane. Many pilots start the engine and head out to the runway, giving no thought to how to use the plane's controls to increase aircraft-handling safety during high-wind or crosswind conditions. Under the right circumstances, a plane can be rocked up onto a wingtip, or worse, flipped over on its back, due to strong winds. In many cases this could have been avoided if the pilot had held the controls in the proper position to compensate for the wind's effect.

Figure 8-1 illustrates how controls should be positioned based on the direction the wind is from. This is accurate for both tricycle- and conventional-gear planes, with one exception: for tailwheel planes, the elevator control should be held in the full aft position when the plane is headed into the wind, as opposed to the neutral elevator positioning in-

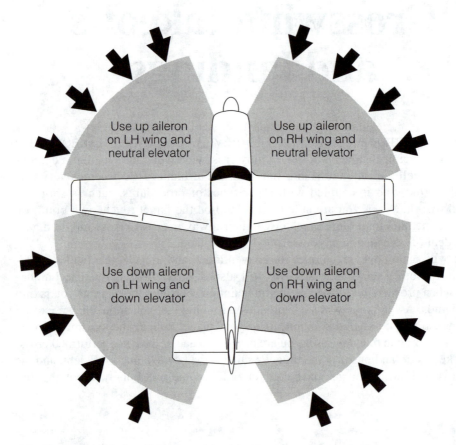

Use up aileron on LH wing and neutral elevator

Use up aileron on RH wing and neutral elevator

Use down aileron on LH wing and down elevator

Use down aileron on RH wing and down elevator

Fig. 8-1. *Crosswind taxiing control positioning.*

dicated in the diagram. When taxiing a taildragger with the wind, the elevator control should be in the forward position. With the controls positioned in the manner shown, the wind will have less tendency to raise the upwind wing and will help to keep it down.

As you can see in the figure, when the wind's direction is off the nose of the plane, ailerons are turned into the wind. This results in the upwind wing's aileron being raised, which helps force the wing down and reduces the ability of the wind to raise it. However, when the wind is from the rear of the plane, the use of ailerons is reversed. In tailwind taxi situations, turn the aileron controls away the wind. This lowers the upwind wing's aileron and helps to force the wing down. When you are taxiing, you will need to be aware of the wind's direction and change the controls to the correct position as you turn the plane.

The elevator control is also used to maintain better control of the plane when taxiing in the wind. For tailwheel aircraft, keep the stick in the full aft position when the wind is off the nose. This helps to keep the tail of the plane down and reduces the tendency for it to come up and the plane to nose over. When taxiing in tailwinds, you must use forward elevator control, though. This again helps to force the tail down as the tailwind flows over the elevators from behind. Like the ailerons, you will need to reposition the elevator control as the plane's direction relative to the wind changes while you taxi. Be sure to avoid keeping the control opposite the correct setting, or the plane's tail could rise very quickly. More than one tailwheel pilot has bent the propeller when the wind has caught them unprepared during taxiing.

When taxiing perpendicular to the wind, both types of planes will have a tendency to "weathervane," or turn, into the wind. Tailwheel aircraft are more susceptible to this, and you must constantly control the plane as you taxi it. This is done with a combination of brakes and steering, but to a large degree relies on use of brakes. When taxiing tailwheel planes in strong winds, use slower taxi speeds, especially when the winds are from the rear of the plane. Slower speeds make it easier to overcome the weathervaning tendency with brakes and steering, since there is less momentum involved.

Be very cautious when taxiing conventional-gear planes in windy conditions. The old adage "Fly a taildragger until the prop stops turning" exists for very good reasons. I have had situations where the wing is lifted by wind alone during taxi at slow speeds, even with the controls in the correct positions. When I worked the flight line, there were occasions when we walked a local pilot's J-3 Cub back from the runway, holding on to the wingtips to keep the wheels on the ground. Sometimes the wind is just too strong, and you will not be able to maintain control of the plane as you taxi it. Once you no longer have sufficient airspeed for flight controls to work, you might be able to land in winds that you cannot safely taxi in. Do not become complacent once the plane is on the ground, whether it is tricycle or tailwheel. When taxiing in windy conditions, use proper control positioning and slower taxiing speeds, and be aware of the wind direction relative to the plane and obstacles near your path. If the plane does weathervane, you want to be clear of anything (like other planes, taxiway lighting, or buildings) that might damage the plane.

CROSSWIND CONTROL USE

Control positioning plays a vital role during crosswind takeoffs and landings. From the moment you turn onto the runway for takeoff, you must manage the control inputs correctly to maintain control of the plane. The same is true of crosswind landings. As you turn onto final, you will be crabbing or slipping the plane to maintain a track parallel to the runway. After touchdown and during the rollout, you must still use the correct control inputs to overcome the effects of the wind. In this section we will cover the use of controls to counteract the crosswind and the rationale behind their use.

In the previous section we discussed control use while taxiing the plane to compensate for the effects of the wind. Many of the principles from that section will also apply to the following discussion of control use during crosswind takeoffs and landings. Let's begin with the takeoff.

As you turn onto the runway, assuming you are taking off into the wind, you will want to roll the aileron control into the wind at maximum deflection. As with taxiing, this positions the upwind wing's aileron in the up position, which helps hold that wing down because as the wind flows over the up aileron, it deflects the wing down. This is true of both high-wing and low-wing planes, and conventional- and tricycle-gear aircraft as well. After you apply power and begin to accelerate on the takeoff roll, the controls will begin to become more effective. As airspeed increases, you might need to reduce the aileron input to prevent the downwind wing from rising. The full aileron input that you start the takeoff roll with is effectively an aileron roll into the wind. The idea is to have just enough aileron input to counteract the wing-lifting force of the crosswind and maintain a wing's level attitude. After the plane lifts off, you will transition to a crab into the wind to maintain a track parallel to the runway heading. Depending on the strength of the wind, it might be necessary to continue to use some aileron input to counteract the wind's effects after lift-off.

Rudder will be used during the takeoff roll in conjunction with the ailerons to maintain the correct track down the runway. As with taxiing, the plane might have a strong tendency to turn into the crosswind as you roll down the runway, and the use of rudder and brakes might be necessary to prevent this from happening. In some cases the weathervaning tendency can be very strong, and a great deal of braking, rudder, and steering are necessary. If you do not remain parallel to the runway during the takeoff roll, the plane might start skidding as the plane changes the direction it is pointing, but momentum continues to carry it on the original heading. When this happens, you can frequently hear the tires squeal as they scrub along the runway. This causes a great deal of tire wear and makes the plane difficult to control.

In some severe crosswind situations, no amount of aileron and rudder can overcome the effects of the wind during the takeoff roll. When this is the case, several things can happen if you attempt to take off. First, the crosswind can cause the upwind wing to be lifted far enough off the ground that the downwind wing is forced into the runway. This can result in anything from a damaged wingtip to groundlooping the plane, or, if the wing digs into the runway hard enough, the plane flipping on its back.

150

Second, the plane can weathervane into the wind. The resulting change in direction can make the plane skid across the runway. This skidding can cause a wing to rise, with the potential outcome the same as was just described. With retractable-gear planes, the resulting side load can also cause the gear to collapse. Third, the plane might be blown across and off the runway into obstacles such as runway lights, other aircraft, or drainage ditches. Obviously, if you attempt to take off in these conditions, you have exceeded the crosswind capabilities of the plane. No amount of crosswind control can overcome the effects of the wind in these conditions, and safety considerations dictate that you not fly until conditions improve.

Control use is much the same after the plane touches down during landing and is rolling out. Ailerons should be rolled into the wind after touchdown to help prevent the upwind wing from rising as the plane rolls along the runway. Depending on the control authority of the ailerons at touchdown, you might not be able to roll in complete ailerons, or the downwind wing might rise. As the plane slows and aileron authority is reduced, you should roll in progressively more aileron control to help keep the upwind wing on the ground. Many pilots fail to do this once on the ground, often neutralizing the controls as they roll out. At this point the crosswind can cause the upwind wing to rise or the plane to weathervane, with the same results as have been previously discussed. Always use crosswind control inputs when the plane is moving.

Rudder should be used during rollout to help keep the airplane parallel to the runway centerline. As with takeoff, the plane will want to turn into the crosswind during the landing rollout. This is very true of tailwheel airplanes, which might require the use of rudder, brakes, and steering to maintain the correct track. One of the difficult problems with crosswinds is that they are seldom constant, and often gusty, so you will constantly need to adjust not only your rudder but also aileron inputs to hold the plane where you want it as you roll out.

One of the main points you must remember is that you must control the plane during the takeoff and landing roll. Crosswinds present a challenge that can be safely overcome if you use the correct control inputs. Fly the plane from the moment you start the engine, and taxi out to the runway until you shut the engine down. A note about the use of brakes for crosswind control: brakes should be used as a last resort, and then with minimum pressure, especially during takeoff. In addition to greater brake wear, excessive use of brakes reduces the plane's ability to accelerate, increasing takeoff distances.

CROSSWIND AIRSPEEDS

When flying in crosswind conditions, it is necessary to have positive control of the plane. In several previous chapters we have discussed flying the plane near the edge of controllability at slow airspeeds. These slower airspeeds result in reduced control effectiveness and a "softer" feel to the controls. Given the need for additional control, it is necessary to use higher airspeeds when flying in crosswinds. The additional airspeed provides a margin of safety and controllability. If standard airspeeds were used for takeoffs and landings in crosswinds, it is possible that even full deflection would not be able to overcome its effects. For this reason you should allow the airplane to accelerate to a higher airspeed during the takeoff roll, prior to rotation. Then, as you rotate and lift-off, you will have more control authority, which allows you to maintain control of the plane during the climb.

The additional airspeed also provides a larger buffer against stalling the plane as you climb. As you are already aware, a plane can stall at any airspeed. The stall is based on the plane's wing reaching the critical angle of attack, which was covered in chapter 2. Changes in wind direction, such as gusts from below the plane, can cause it to stall even though no change in pitch has taken place. You might have noticed that as you climb on turbulent days, the plane's stall-warning horn will sound briefly as the plane is buffeted by gusts. This is due to the rapid change in wind direction that takes place as you fly through the turbulent air. The additional airspeed you allow the plane to fly at not only provides a safety margin to keep you out of a potential stall, but also helps you maintain control of the plane as you encounter these gusts.

Landings in crosswind, gusty, or turbulent conditions also require that you carry some additional airspeed during the approach to provide better control of the plane. This is done for the same reasons that were mentioned for takeoff. The extra airspeed gives you greater control authority and a larger margin against stalls caused by gusts or turbulence. You will probably also want to touch down at slightly higher airspeeds to maintain this control through the flare and touchdown. When flying in strong crosswinds or gusty conditions, I prefer to do this as opposed to full-stall landings because of the extra control it gives. In full-stall landings, you will have minimal ability to overcome the effects of a crosswind, such as a wing being raised or the plane weathervaning. In essence, you are flying the airplane down to the runway, which helps you maintain control of the plane throughout the landing. In later sections of this chapter, we will discuss this in more depth, especially as it is related to tailwheel landings.

In many cases it is also advisable to carry some engine power down to, or just prior to, touchdown. I have found this provides more positive airspeed control and smaller fluctuations in glideslope. Depending on the plane, you might need to reduce power to idle prior to actually touching down. When you do this, be prepared to compensate for the change in pitch the power reduction might cause. Be sure to reduce the engine to idle power and avoid leaving power in after touchdown, which will increase the landing rollout.

There is no equation or table that I am aware of that indicates how much additional airspeed is necessary during the takeoff and landing to provide an acceptable margin of safety. I have been told that you should add the equivalent of the crosswind component to normal airspeeds, but I have not been able to document this statement. I normally add enough airspeed to give the airplane the controllability that lets me get the airplane back to level flight after gusts without excessive control deflection. This can vary, depending on the given conditions I am flying in, the type of aircraft I am flying, and how the plane is loaded. You will need to develop a sense for the correct airspeeds you should use for your plane. In higher crosswinds or more turbulent conditions, I use a higher airspeed than in less severe situations. You must decide the correct airspeed for a given condition, though. If you use an airspeed higher than conditions warrant, you will use more runway than necessary and cause additional wear on the tires and brakes. An airspeed that is too low will result in lack of aircraft control. The more experience you have flying your plane in crosswind conditions, the easier it will be to judge the correct airspeed to use. If the manufacturer has recommended rotation or approach speeds for given crosswind conditions, use those.

One very important point to remember is not to exceed the demonstrated crosswind component for your airplane. This is the maximum crosswind the plane was tested in and is certified to safely land in. This value is documented in the plane's operations manual and requires the correct crosswind procedures to fly the plane in those conditions. If you attempt to fly in excess of the maximum crosswind component, it is possible that full-control deflection will not be able to overcome the effects of the wind. Keep in mind that just because a plane is certified to land in a given crosswind component, if your technique is rusty or incorrect, you might not be able to take off or land safely. In chapter 6, I relayed my aborted attempt to land a taildragger in a strong crosswind.

I don't recall what the crosswind component was that day. It is very likely the plane was up to the task of landing, but my level of crosswind proficiency at that time was not. This is where practice becomes very important. The more you practice landing in crosswinds, the better you will become. If you are at all rusty, or the conditions are stronger than you have flown in previously, be sure to get a qualified flight instructor to ride along with you. Make sure you practice, though. Almost any day there is an airport somewhere nearby that has a crosswind. I actually prefer to land the Pitts Special I fly in a crosswind because I can see the runway ahead somewhat easier over the wing lowered into the wind. Remember that safety is your highest priority, and if conditions are beyond the plane's or your capability, do not fly until conditions are less severe. Maintain adequate airspeed during takeoff and landing to give you the necessary control, and do not exceed the demonstrated crosswind component for your plane.

TAKEOFF

In this section we will discuss crosswind takeoff procedures. This will follow the normal format of topics: flap settings, trim setting, and procedures for tailwheel and tricycle-gear planes. The purpose of crosswind takeoffs is to safely compensate for the effects of crosswinds during the takeoff roll, rotation, and climb. We have already discussed some of the potential hazards that you can encounter during crosswind takeoffs, such as the upwind wing being lifted or the plane weathervaning into the wind. The correct crosswind procedures will help you maintain control of the plane and avoid these situations.

I enjoy flying in crosswind conditions and feel it is an important part of the curriculum for any pilot. Much to my students' dismay, I have gone out of my way to find crosswind conditions for them to practice in. As you gain more experience in crosswind takeoffs, you will become more confident in your abilities and safer while performing them. Many pilots forget to use the proper crosswind procedures during takeoff. While they might get the plane off the ground, the takeoff quite often could have been done more smoothly and more safely. As you read the following sections, think about how the procedures affect the plane as it takes off. The better your understanding of how you fly crosswind takeoffs, the more automatic that compensation will become.

Flap setting

Use of flaps in crosswind takeoffs requires an understanding of the plane you are flying and the conditions you are flying in. During takeoffs, flaps reduce the stalling speed, allowing the plane to leave the runway at a lower airspeed. We have already touched on the need to maintain adequate control of the plane in crosswind operations and the need for additional airspeed to do this. If flaps allow the plane to fly at a slower airspeed, then in crosswind flying their use can make the plane less able to overcome the effects of the crosswind, even when you use proper crosswind control inputs. Incorrect use of flaps can be counterproductive in crosswind situations, and in this section we will discuss some of the considerations you should make when using flaps in crosswind takeoffs.

The effects of flaps vary from plane to plane. I have flown some trainers on which flap extension only reduced the stalling speed by a few knots. On these planes, the flaps seemed to be there only to give the student practice in the use of flaps, with very little stall-speed reduction. On other planes the stall speed can be reduced a great deal when the flaps are extended. The latter plane might present more of a control problem with fully extended flaps than the plane on which flaps make little difference in stalling speeds. Given the same crosswind conditions, the pilot of the first plane might be able to use flaps during takeoff and have no crosswind control problems, while the pilot of the second plane might experience loss of control when using flaps.

You need to understand how the use of flaps affects your airplane's controllability during takeoffs and apply this knowledge when making crosswind takeoffs. Depending on your plane, it might be necessary to reduce the manufacturer's recommended flap setting when making a crosswind takeoff. If a specific setting is recommended for a normal, short- or soft-field takeoff, it might be necessary to reduce the setting, or avoid the use of flaps, to maintain the necessary control during the takeoff. The need for reduction of flap use can vary, based on the value of the crosswind component. In light crosswinds, standard flap settings might be possible, while in moderate crosswinds you might need to reduce the use of flaps by one notch. In strong crosswinds, near the limit of the maximum demonstrated crosswind component, you might find it necessary to forego the use of flaps completely.

When it is necessary to reduce or eliminate flaps during crosswind takeoffs, keep in mind that this might increase the takeoff distance. Crosswind takeoffs can require a longer runway than indicated in the takeoff performance charts for your plane. Some performance charts indicate additional takeoff distance requirements with differing flaps settings, while others do not. In these situations, your experience flying the plane will come into play as you estimate runway-length requirements. In questionable situations, err on the side of being conservative in your estimates.

The best way to understand the correct flap settings to use during takeoff is to practice them. Find a sufficiently long runway and try takeoffs beginning with normal flap settings, then reduce them incrementally until you are taking off with no flaps. Initially do this without a crosswind, then with small crosswinds, working upwards in crosswind intensity. This should be done for normal, short- and soft-field takeoffs. As you gain experience with the plane in these situations, you will also be building judgment capabilities. With sufficient practice, you will be able to judge how your plane will react in a given crosswind takeoff. Knowing this before the takeoff begins is the key to successful takeoffs.

Trim setting

Use the manufacturer's recommended trim settings during most crosswind takeoffs. This provides the correct elevator pressures during the takeoff roll and initial climb. For tailwheel planes I have flown, this holds especially true. Incorrect trim settings can cause the tail to rise too early, or require excessive forward stick to get the tail off the ground during the takeoff roll. Even with tricycle-gear planes, you should use the correct trim settings to prevent the need for excessive back pressure to raise the nose of the plane. Correct trim settings also prevent the nose from rising too early and the plane from flying at low airspeed. On some tricycle-gear aircraft, due to the need for additional speed prior to rotation, I have used a very small amount of nose-down trim to help keep the plane on the runway until adequate airspeed has been achieved. When trim is set to standard settings, certain planes might have a tendency to fly off the runway as you allow them to accelerate above the normal rotation speed. This additional airspeed can be helpful in maintaining control in strong crosswinds.

There are complications with doing this, though. First, excessive nose-down trim causes wear on the nosegear and tire. In some instances, violent shimmying can take place as a result of this. Excessive nose-down trim might require a great deal of elevator back pressure to rotate once the correct speed has been reached. The plane might also have a tendency to nose back down to the runway after lift-off. If you use nose-down trim, make sure only a very slight amount is used.

Conventional-gear ground roll and rotation

In this section we will discuss crosswind ground roll and rotation procedures. I will be using two examples, a Citabria and the Pitts Special S-2B. These aircraft exhibit different tendencies during crosswind takeoffs and serve as an example of contrasting techniques. As you read this section, keep in mind that the procedures might need to be modified slightly for your plane and a given set of conditions.

As you turn onto the runway, roll full ailerons into the wind. We have previously established that this reduces the wind's ability to lift the upwind wing. Add power normally, and as the plane accelerates, adjust the aileron input to correctly compensate for the amount of crosswind you are experiencing. The right amount of aileron input will be different, based on each set of conditions you are flying in. The purpose is to prevent the upwind wing from rising and to help keep the plane from skidding sideways across the runway. Rudder, tailwheel steering, and brakes might be necessary to keep the plane's longitudinal axis parallel to the runway, depending on the strength of the crosswind. In most cases, you will want just enough aileron input to keep the wings level during the takeoff roll. When crosswinds are very strong, keeping the wings level during the takeoff roll might not be sufficient to overcome the "skipping" tendency that can develop as the plane rolls down the runway. Strong crosswinds can cause the plane to be pushed sideways by the wind, even while the plane is still on the runway. This can cause the landing gear to skip as it is forced sideways. In this case, it might be necessary to hold enough aileron into the wind to raise the downwind wing and main landing gear to prevent the plane from continuing to skip across the runway. As aileron is added, opposite rudder is input to keep the plane's longitudinal axis parallel to the runway. Figure 8-2 demonstrates this technique. As you can see, the plane can be balanced on just one of the main landing gear. By banking into the wind during the takeoff roll, the plane can be stabilized to roll straight along the runway.

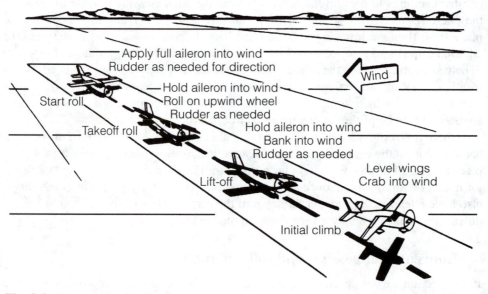

Fig. 8-2. *Crosswind takeoff roll.*

Due to the varying speed of a crosswind, you will need to monitor the plane and adjust your control inputs accordingly to just overcome the effects of the crosswind during the takeoff roll. Let the plane accelerate to an appropriate airspeed while still on the runway, then rotate as you normally would. Establish a positive rate of climb as quickly as possible and climb out. Once clear of the ground, you should establish a crab into the wind to allow you to continue tracking the runway centerline until you make your first turn in the pattern.

These general techniques hold true for both three-point and wheel takeoffs in conventional-gear planes. If you perform wheel takeoffs, you might experience a stronger turning tendency into the wind as the plane's tail is raised from the runway. With a steerable or locked tailwheel, a certain amount of directional control is lost as the tailwheel leaves the runway. This can normally be compensated for with the use of rudder. When the crosswind is strong enough that the previously mentioned wing-raising techniques are required, use caution. In biplanes and low-wing taildraggers it might be quite possible to drag the upwind, lowered wing as you roll along prior to leaving the runway. High-wing planes have more distance between the wing and runway, but it is still possible to overcompensate for the crosswind and drag the wingtip on these planes as well. There might be less stability as the plane rolls along on only one main landing gear, and you will need to make sure the plane does not get out of control. When flying in crosswinds that are this strong, you should be very proficient and familiar with the plane you are flying. This forward slip should be maintained through rotation and until well clear of the ground. At this point the plane should be transitioned from the slip to a crab into the wind to maintain a ground track parallel to the runway.

In the Citabria I fly, standard crosswind takeoff techniques work well. I start the takeoff roll with the ailerons rolled fully into the wind. As the plane accelerates, the ailerons are adjusted to keep the wings level, and rudder is used to maintain the correct orientation with the runway centerline. When adequate speed has been achieved, I raise the tail and continue to accelerate on the main landing gear. The additional airspeed provides greater control authority, often making it necessary to reduce the amount of aileron input. As the Citabria lifts off, I then crab into the wind to continue to fly along the runway centerline.

The Pitts S-2B requires a somewhat modified version of these techniques. Due to its control authority and sensitivity, the Pitts requires slightly different control inputs that begin with little or no aileron rolled into the wind as power is applied. With four ailerons and lots of acceleration, it is easily possible to roll the plane up on a wingtip with too much aileron input. Rudder use is normal, with just enough to keep the plane's longitudinal axis parallel to the runway. In normal crosswinds, I like to start out with little or no crosswind aileron correction and add it as necessary during the takeoff run. In stronger crosswinds I begin with a small amount of aileron into the wind and adjust it accordingly. With its stubby wings and aileron spades hanging down below the lower wing, the Pitts is less tolerant of overcontrolling the aileron input during the takeoff roll. Like the Citabria, I raise the tail at the appropriate airspeed and accelerate to rotation speed on the mains. With the rate of climb the S-2B has, it gets away from the ground very quickly. I then crab into the wind to maintain the correct ground track.

The largest percentage of errors committed by most pilots during crosswind takeoffs is the failure to roll ailerons into the wind before the takeoff rolls begin. They start with the ailerons neutral and keep them in that position during the entire takeoff roll. While it is possible to get away with this technique in weak crosswinds, it is a very bad habit to get into. First, it demonstrates a lack of understanding of proper crosswind takeoff procedures on the part of the pilot. If pilots do not understand how crosswinds can affect them, they are less concerned with correcting for them. Second, bad habits that are developed and reinforced through repetition become second nature for pilots. When the situation becomes more critical, these unsafe techniques become the automatic reaction when pilots encounter strong crosswinds. Many pilots also fail to compensate for the weathervaning tendency of the plane. This becomes an even larger problem as the plane's tail is raised to get it up on the main landing gear. I have seen pilots raise the tail, then start heading toward the side of the runway as the wind changes the direction the plane is traveling. Instead of steering the plane back to the correct heading, they attempt to become airborne before they reach the side of the runway. Be prepared for the potential direction change and react to compensate for it. Attempting to lift off at too slow an airspeed is a problem some pilots also have. This reduces their ability to compensate for the crosswind. Finally, many pilots do not establish an adequate crab into the wind after takeoff, allowing the plane to drift off the runway centerline during the initial climb.

Tailwheel airplanes are not inherently more difficult to fly than tricycle-gear planes in a crosswind takeoff, but they do demand that you, the pilot, fly them at all times. If you let the wind control the plane during the takeoff, using the same takeoff techniques you do on a calm day, the outcome will eventually be one you are not looking forward to.

Tricycle-gear ground roll and rotation

Tricycle-gear planes employ many of the same crosswind takeoff techniques that are used on conventional-gear planes. There are those pilots who feel the nosewheel steering found on many tricycle-gear planes provides greater control during crosswind takeoffs. Do not, however, be lulled into a sense of complacency by this very common belief. If you use improper crosswind takeoff techniques when flying a tricycle-gear plane, it is just as likely to surprise you during takeoff as any tailwheel plane. In this section we will cover crosswind takeoff techniques for tricycle-gear planes, comparing them to conventional-gear planes in the process.

For all tricycle-gear planes I fly, I roll full aileron into the crosswind as I turn onto the runway. Unless you happen to be flying a tricycle-gear plane that is similar in control authority to the Pitts Special, this is probably the way you should begin all crosswind takeoffs. The elevator should be in a neutral position, and the rudder should be centered as you apply takeoff power. Like tailwheel planes, as the plane accelerates you might need to reduce your aileron input to maintain wings level during the takeoff roll.

Tricycle-gear planes exhibit a weathervaning tendency, like taildraggers, that you must compensate for. The severity of this tendency will vary among different aircraft. In many cases, nosewheel steering easily overcomes the tendency to turn, while in other cases weathervaning is very pronounced. It has been my experience that T-tail aircraft are more susceptible to weathervaning. This is due to the additional rudder and elevator area the crosswind can get under and push against during the takeoff roll. In a strong crosswind, the nosewheel might scrub along if you use only steering to keep the plane straight during the takeoff roll. In these situations it might be necessary to use sufficient aileron and rudder to lower the upwind wind and let the plane accelerate with the down-wind main gear off the runway. This sideslip during the takeoff run was discussed pre-viously in this chapter in the section on tailwheel, crosswind takeoffs. Figure 8-2 depicts the use of controls during this type of takeoff.

Once the plane accelerates to the desired airspeed, rotate and establish a positive rate of climb. After you clear the ground, transition to a crab into the wind to maintain the correct ground track. As you can see, there is very little difference between con-ventional and tricycle crosswind takeoffs.

Common mistakes pilots make during tricycle-gear takeoffs include all the points mentioned previously for tailwheel planes. One that is unique to tricycle-gear planes is, in strong crosswinds, compensating for the crosswind with nosewheel steering alone during the takeoff run. This causes a great deal of wear on the tire as it scrubs along. When proper techniques are used, this reduces wear on tires and provides better con-trollability. One other point to remember is that retractable-gear planes are susceptible to the gear collapsing when heavy side-loads are imposed on them. Letting the plane drift sideways across the runway places potentially heavy side-loading on the landing gear.

Before leaving this section and moving on to crosswind landing procedures, we need to discuss one last topic associated with crosswind takeoffs. If the crosswind is perpendicular to the runway, you can help reduce its effects with the direction you choose to use taking off. We have established that various forces (such as P-factor) act-ing on the plane during takeoff result in a left-turning tendency for the plane. Use of right rudder is needed to counteract these forces, but a crosswind perpendicular to the runway can aggravate this if you take off with the wind on the left. This is because the plane has more fuselage area behind the landing gear, resulting in the plane weather-vaning or turning into the wind. A direct crosswind from the left causes the plane to turn to the left, adding even more left-turning tendency to the ones normally present. By taking off with a direct crosswind on the right of the plane, you reduce the effects it can have on the plane during takeoff, and the amount of control input that will be necessary as a result of it. Now let's move on to crosswind landing procedures.

LANDING

Crosswind landings require that you know the proper crosswind procedures, correctly apply them during the approach, and continually monitor the adjustments necessary to maintain the desired ground track in what are usually varying crosswinds. For student pilots, crosswind landings require a great deal of concentration and focus to perform ac-curately. In many cases you might fly a 1- or 2-mile final during the approach, and this will require you to adjust for the crosswind to maintain the correct ground track while still well out from the runway. As you transition through the flare and touch down, you must maintain your crosswind control inputs to keep the plane heading in the right direction down the runway. Without those control inputs, many of the same hazards that we discussed for crosswind takeoffs are possible. In this section we will review the correct crosswind landing procedures for both conventional- and tricycle-gear planes, beginning from the time the plane is turning to final through touchdown and rollout.

I like making crosswind landings. Understanding what the wind is doing to the flight path of the plane and correcting for it becomes almost second nature after you become accustomed to the correct procedures. It's a real sense of accomplishment to smoothly land and roll out when crosswinds are at the limits of the plane's capabilities. Practice is the key to well-executed crosswind landings, and the more you do, the better you will become. Don't just jump in your plane and head off into crosswinds, though, unless you are proficient in your techniques. Find a qualified flight instructor to help you become proficient if you are at all unsure or rusty.

Flap setting

We have previously discussed how flaps affect the plane during crosswind takeoffs. In this section we will review how they can affect landings in crosswinds and what you should consider when deciding which flap setting should be used for the landing.

During landings, flaps reduce the stall speed of the plane, allowing slower approach speeds and shorter landing rolls. In landing configurations, flaps also increase drag, which allows you to maintain a steeper glideslope during approach. In the crosswind takeoff flap discussion, we talked about the need for additional airspeed taking off, which provides additional capability to overcome the effects of crosswind. As with the takeoff, insufficient airspeed during the approach and flare might not give you enough control authority to compensate for the crosswind. Maintaining a higher airspeed during the approach and landing will help provide better control to maintain the plane's attitude.

As you decide which flap setting to use, you need to understand how that will affect your ability to control the airplane close to the runway. Much of what we reviewed in the crosswind takeoff section holds true for landings. In this section we will cover some points specifically related to crosswind landings.

As crosswinds increase in strength, it becomes necessary to maintain a higher airspeed during the approach. Airspeed can be kept higher, even with full flaps extended, during the final approach leg of the landing, up to V_{FE}. But as you flare the plane and attempt to touch down, there might be a greater tendency to float as excess airspeed bleeds off before touching down. During this period, the plane's controls become progressively less effective, and strong crosswinds can become a problem.

Using less flaps in crosswinds results in being able to get the plane planted firmly on the runway with less potential for floating and more control effectiveness. This is, of course, dependent on good airspeed control during the approach and flare. Smaller flap settings also allow you to more easily maintain a higher approach speed on final without resorting to very steep descent angles.

Experience with the plane you fly will help you determine what the correct flap settings should be. Your aircraft's manufacturer might also have recommended flap settings for crosswind landings. Also remember that higher approach and landing speeds will likely result in longer landing rollout lengths. As you decide what flap setting to use for a given crosswind situation, you will need to choose what the most favorable balance should be between crosswind strength, approach speed, and runway length available.

With light crosswinds, you might be able to use normal flap settings and maintain adequate control throughout the approach. In moderate crosswinds you might still be able to use standard flap settings, but you might find it less work to control the plane by reducing them a notch or so. In very strong crosswinds, only one notch or no flaps at all might provide the best control feel during the flare and landing. My experience has reenforced this pattern of flap use when I fly in crosswinds.

Practice crosswind landings with varying amounts of flaps to find out what works best for your plane in a given crosswind situation. As you gain experience, you will be better able to judge what the correct flap settings should be for crosswind conditions. Your plane might react differently than those that I have flown, so use the previous information as a baseline to work from. One note: do not add flaps during the approach and retract them prior to touchdown in an attempt to increase the stall speed during or after the flare. The resulting pitching moment and speed change might cause the plane to pitch up and stall while still well off the runway. The reduction in drag from the flaps might also cause the plane to increase airspeed, and the reduction in lift might cause the rate of descent to increase dramatically, causing the plane to impact the runway at a high rate of descent.

Trim settings

Use trim settings that result in the plane stabilizing at the correct approach speed during crosswind landings. If you are using higher speeds during the approach, setting the trim early will help you maintain the correct value during the rest of the approach. Sometimes pilots slip back to "standard" approach speeds out of habit during a crosswind approach. In this case, having the plane trimmed for the desired airspeed can help overcome this tendency.

Conventional-gear flare and landing rollout

In this section we are going to review the approach, flare, and landing for conventional-gear planes. During each of these phases of crosswind landings, you must be aware of the effect the wind is having on the plane and compensate for it. Depending on how far out you are from the runway after the turn to final, what might seem like small deviations from the runway centerline can become very large if left uncorrected. During the flare and rollout, the plane can be blown off the runway if you do not monitor the airplane and what the crosswind is doing to it.

Two maneuvers play a major roll in crosswind landings. These are the crab, used during final approach, and the sideslip, helpful during the flare and touchdown. In Fig. 8-3, a plane on final approach is crabbing into the wind to maintain the correct ground track in relation to the runway's centerline. A crab is achieved by adjusting the plane's heading into the direction the wind is coming from. The turn into the wind and ongoing crab are flown in a coordinated manner. The amount, or angle, of crab necessary will change based on your airspeed, the strength of the crosswind, and its angle compared to the runway and the desired ground track. Correctly performed, a crabbed airplane flies into the wind enough to offset the pushing effect the wind is having. The result is a straight track over the ground to the runway along its centerline.

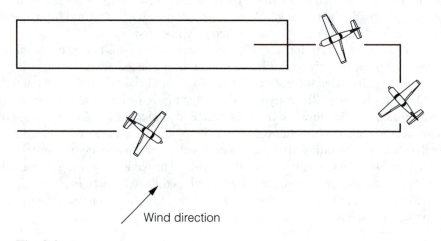

Wind direction

Fig. 8-3. *Crosswind correction in pattern.*

The sideslip is illustrated in Fig. 8-4. This is useful during short final approach and flare. Properly executed, it allows the plane to maintain its longitudinal axis parallel to the runway, yet prevents the crosswind from blowing the plane across the runway. The sideslip is entered by lowering the upwind wing into the wind and applying opposite rudder to keep the plane's longitudinal axis parallel to the runway's centerline. The sideslip prevents the crosswind from blowing the plane laterally across the runway, yet the plane's longitudinal axis remains straight down the runway. How much you need to lower the upwind wing depends on the angle and strength of the wind. The stronger the wind, or the greater the angle to the runway, the more you will need to lower the wing into it. This, in turn, affects the amount of opposite rudder needed to keep the airplane straight. As you lower the wing more, you will also need to apply additional opposite rudder.

There are limits to how much sideslip you can use, though. In strong crosswinds it is possible that you would need to lower the upwind wing to such a degree that the wingtip strikes the ground. In this case, you are probably well beyond the crosswind component limits of the plane. Low-wing planes are somewhat more susceptible than high-wing planes, but it is possible for both to be damaged in this manner.

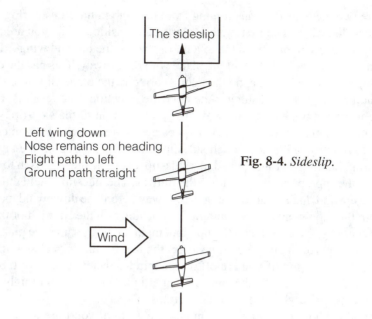

The sideslip

Left wing down
Nose remains on heading
Flight path to left
Ground path straight

Fig. 8-4. *Sideslip.*

Wind

Let's begin our approach after we have turned onto final, about 2 miles out from the runway. All of the landing principles that we have covered, such as airspeed control and touchdown point targeting, still apply for crosswind landings. As you turn onto final, you should automatically adjust your turn, whether it was made to the left or right, so that the plane is crabbed slightly into the wind. As soon as the turn is completed, you should monitor your track to the runway. If you are drifting, alter the amount of crab (Fig. 8-3) to overcome that movement. If the wind is pushing you away from the runway centerline, increase the crab angle. If you are drifting into the wind, reduce the crab angle. Monitor your track to the runway as you continue your approach on final. If the crosswind varies in its intensity, angle, or both, it will be necessary for you to continually adjust the crab to maintain the correct track. Remember that you will also need to use an airspeed that gives you adequate control. Airspeed control becomes even more important in turbulent conditions due to the rocking motions turbulent air might cause for the plane. Soft, slow controls will make it more difficult to return the plane to the proper attitude.

As you get closer to the runway, it becomes easier to judge your drift from the runway centerline. Continue to adjust your crab angle as necessary to stay on the correct ground track and watch your touchdown point to verify that you have the correct glideslope. Depending on the strength of the wind, it might be necessary to begin your final approach descent closer in to the runway. Due to the slower ground speed that is possible with a strong headwind/crosswind, if you start your descent at the normal point, your glideslope might not carry you to the runway. As you watch your touchdown point, you will be able to judge what the necessary correction is.

There are two methods for transitioning from the crab to the sideslip (Fig. 8-4) as you get in on a close final. The first method starts the transition while you are on short final. At approximately ¼ mile from the touchdown point, the upwind wing is lowered and opposite rudder is applied to align the plane with the runway. This has the effect of the plane slipping into the wind, and when done correctly the plane slips into the wind the same amount the plane is pushing it the opposite direction. The result is the plane flies a straight ground track to the runway. As you transition to the sideslip, you will need to monitor and maintain the correct airspeed and ground track. With practice you will be able to roughly judge how much sideslip is required, based on the amount of crab used during final. Once transitioned to a sideslip, continue the approach to the runway, flaring at the appropriate height above the runway. The sideslip is held through the flare, with the upwind, main gear touching the runway before the downwind main. This is true of both three-point and wheel landings. Depending on the strength of the wind, it might be necessary to roll along on the upwind, main landing gear as the plane slows.

This sideslip approach gives you time to get the feel for the correct amount of slip to correct for the crosswind. If you are off on the initial sideslip, you have time to adjust it to maintain your track to the runway centerline. Once set, you might need to make adjustments for varying crosswind directions or speed.

The second method involves maintaining the crab until you flare the plane, then transitioning to the sideslip during the flare. With this method you fly the crab throughout the approach, to the point at which you begin to flare. As before, roll the upwind wing down and use opposite rudder to remain parallel to the runway. Depending on when you transitioned to the sideslip, the flare can be entered either during the transition or immediately afterward. Like the first method, the upwind main will touch before the downwind main gear, and this is again true for both wheel and three-point landings.

The techniques for entering the sideslip are the same as for the first method, but they take place later in the approach. This technique is slightly more demanding than the first method in that you have less time to judge the amount of sideslip necessary for the crosswind, and you are combining the control movements for sideslip entry with the flare control inputs. In addition, you now must judge not only your flare height, but also whether the plane is drifting in relation to the runway during the transition. When using this method, you must know the plane you are flying and be experienced in crosswind landings to accurately judge the amount of slip to be used initially.

One difference between wheel and three-point landings is the visibility they offer you ahead for judging drift. Visibility over the nose in wheel landings should provide a clear view of the runway to aid you in determining if there is any drift. When making three-point landings, visibility ahead might be reduced during the flare, necessitating the use of peripheral vision to determine if there is any drift.

Method two reduces the amount of time spent in the slip, which, depending on the plane you are flying, can feel abnormal due to the "crossed" control inputs. There is also an increased chance for a stall/spin situation to develop with "crossed" controls if the airspeed is not properly managed during the sideslip. Method two reduces the potential for this to take place.

With both methods, as the controls lose effectiveness with slower speeds, gently lower the downwind, main landing gear to the runway. This should be done before there is no aileron authority and the raised main wheel drops uncontrolled to the runway. Once the mains are on the ground, if your plane will not overrespond, roll full aileron into the wind to help keep the upwind wing from being raised, as previously discussed. Otherwise, feed enough aileron into the wind to keep the wing from being raised. If you are performing wheel landings, lower the tail in the normal manner as the plane slows. After the tail is on the ground for both three-point and wheel landings, keep the control stick fully aft. At some point you will slow enough that full aileron should be rolled into the wind. Also, be prepared for the weathervaning tendency that might be present. As the plane slows, additional rudder, braking, and tailwheel steering will be necessary to compensate for the plane's turning tendency. Once the plane has slowed enough, use the crosswind taxi techniques we have already covered.

Properly performed, crosswind landings in taildraggers are as safe and easy as any other landing. Common mistakes include not correcting the crab to maintain the correct ground track to the runway and poor airspeed control. Pilots frequently forget to use opposite rudder as they lower the wing for the sideslip. This results in the plane's longitudinal axis not being parallel with the direction of flight, and as the plane touches down, a side load is placed on the landing gear. The plane then tends to swerve as it aligns itself with the direction it is traveling. Finally, many pilots do not continue to use proper crosswind controls after the plane touches down, feeling that once on the ground they can drive the plane to the ramp and not worry about flying it. When flying at the limits of the plane's capabilities, you should be aware of the situation and use the correct techniques. Be sure to fly the plane to the ground and use the controls properly until you shut the engine down.

Tricycle-gear flare and landing

The crosswind approach, except for the flare, is flown exactly the same for tricycle-gear planes as just described for conventional-gear planes. The crab is used on final, and either sideslip method (already discussed) can be used. Use of controls will also be the same as covered in the previous section.

The one difference between the planes is the touchdown attitude. You will want to land in the normal nosewheel-high attitude that is typical of tricycle-gear planes. If necessary, the upwind main gear should touch first, then the downwind main should be lowered in a controlled manner. Gently lower the nosegear to the runway, and use crosswind controls in the normal manner.

You might find, due to the higher approach and landing speeds, that the plane touches down in a flatter attitude, with the nose lower than during other approaches. This more level attitude helps reduce the tendency the plane has to float down the runway. We have discussed on several occasions that the coefficient of lift is related to the angle of attack. With a flatter angle of attack during the flare, the plane will produce less lift as compared to the normal flare attitude. This allows you to plant the plane more firmly, at a higher airspeed than would be otherwise possible.

In a strong crosswind, maintain the sideslip as long as possible after touchdown to avoid having the nosewheel scrub along the runway as the plane is pushed by the crosswind. Once you are slow enough, nosewheel steering will be sufficient to maintain directional control. During the transition period from touchdown to taxi speed, the plane might have a strong tendency to weathervane. Use of the correct techniques can reduce wear on the plane's landing gear and tires.

Many of the mistakes common to conventional-gear planes also apply to tricycle-gear planes. The most common, for both types of planes, is the pilot fails to position the controls in the correct manner after touchdown. Force yourself to do this. Do not get out of the sideslip too early after touching down. It might seem like a difficult balancing act to hold the plane on one main landing gear, but this can add a great deal more control to the landing rollout. Finally, be sure to position the controls properly as you taxi. Like tailwheel planes, fly them until you shut the engine down.

SUMMARY

In this chapter we have covered crosswind takeoff and landing techniques. You should now have a very good understanding of the proper procedures to control the plane in crosswind conditions. We have covered taxiing, takeoff, and landing procedures for crosswind operations. At this point let's consider how it might be necessary to combine crosswind procedures with other types of takeoffs and landings, such as soft- or short-field operations.

When flying into or out of a soft or short field in a crosswind, you will use the same basic procedures that were discussed for those situations, but you must also counter the crosswind you are flying in. When adding a crosswind into the mix, you will need to combine the procedures for crosswind work with the other. During a crosswind soft-field takeoff, for example, you will also need to use the correct aileron and rudder inputs for the crosswind. Depending on the strength of the crosswind, you might also want to increase the speed at which the plane leaves the ground by reducing the amount of back pressure and the angle of attack during the takeoff run. The same type of considerations hold true when crosswinds are part of any specialty takeoff or landing.

Practice pays off in these situations. If you regularly practice and understand the correct procedures for the individual takeoff or landing you are attempting, combining them with crosswind procedures becomes almost second nature. It is quite possible to execute short- and soft-field takeoffs or landings in crosswind conditions and have performance figures be very close to standard.

Know the proper way to execute crosswind takeoffs and landings and practice them as often as you can. Be sure to use the correct procedures for the plane and conditions you are flying in. If you have any doubts, take a qualified flight instructor with you for additional pointers, but be sure to go out and practice.

9
Winter flying

WINTER FLYING IS A TOPIC SOME PILOTS ARE VERY FAMILIAR WITH, WHILE others might have never encountered it. Flying during the winter offers some very scenic views from the air and, very often, glass-smooth air that is free of the turbulence often generated as a result of thermals during the summer. For pilots in the less-moderate latitudes, every year brings the onslaught of winter, and with it, the effects it has on the planes they fly. In this chapter we are going to discuss a number of topics related to aircraft operation during the cold, winter months.

Growing up in the Midwest, I thought everyone understood that there are some special considerations you must make when flying during the winter. However, one December day, as a relatively fresh private pilot, I was approached by an older pilot while preflighting the plane I was getting ready to fly. He asked if the heavy frost on his wings was anything to be concerned about. At that point it struck me that not everyone receives the same education related to winter flying. I explained the potential danger of the frost related to airflow and lift. The pilot thanked me for the help and cleared his wings of the frost before taking off. It is for that reason that a chapter in this book is dedicated to takeoffs and landings and how they are affected by ice, snow, and frost. While it is not possible to cover every conceivable situation, this chapter should be of aid to those who are not familiar with winter flying.

In this chapter we will review a number of winter-flying topics. Among them are aircraft preflight, taxiing and braking, the takeoff run, landing rollouts, and skids. A number of other topics will also be interspersed among these general areas. When you have completed this chapter, you should have a better understanding of the peculiarities associated with winter flying.

PREFLIGHT

Preflights during the winter require you to include a few special steps that are not part of the normal preflight routine during warmer months. The cold weather can cause problems with the plane's engine, empennage, and landing gear. In this section we will cover these special, winter preflight topics, including engine preheating, frost, snow, or ice on the plane, and how snow or ice can affect tires and wheelpants. Our first topic is the engine preheat.

ENGINE PREHEAT

When an aircraft engine is exposed to winter's cold, even for a relatively short period of time, the engine and oil cool at a rapid rate. Most modern aircraft engines are constructed of aluminum for lightness and strength. They also tend to radiate heat at a high rate and, as a result, cool down quickly. In the cold temperatures of winter, the oil in an aircraft engine also cools and thickens much faster than during the summer. After the engine starts, it is crucial for oil to begin flowing as quickly as possible to prevent excessive wear. However, the cold, very thick oil present in the engine can take a great deal of time to warm up and begin flowing through the engine. This increases engine wear and reduces the life of the engine. In severe cases the engine can seize before the engine's heat warms the oil enough for it to begin to properly flow through the passages and channels.

To help prevent the potential damage during the winter, it is common practice to preheat an aircraft engine prior to starting it. The preheater itself can be constructed a number of different ways. In general, though, hot air is blown into the front of the engine cowl, flowing through the engine compartment and warming the engine components. As the metal of the engine warms, so does the oil.

Figure 9-1 shows a typical engine preheater for small aircraft. Normally, a propane or kerosene heater blows hot air into two exhaust pipes, which fit into the front of the engine cowl. Depending on the outside air temperature, it might be necessary to let the heater run for 20 minutes or more to warm the engine to proper temperatures. It is also a good idea to drape a blanket over the top and sides of the cowling to help hold the heat in and insulate the engine compartment from the outside air. I have flown in the cold of Wisconsin winters when the temperature is 20° below zero and, without first preheating the engine, there is no chance it will start. On days when the temperature is that low, if the engine is not started within approximately 20 minutes of completing the preheating process, it will be necessary to preheat it again.

Fig. 9-1. *Engine preheater.*

As you are preheating the engine, there are some things to look for to avoid damage to the plane. Make sure the heating tubing that is run into the front of the engine cowling is not directed at the top or sides of the cowling, but at the engine itself. The air coming out of the heater can become very hot, and, if blown directly onto the cowl, can cause the paint to bubble up. If the cowl is made of fiberglass, the direct heat can also cause the fiberglass to melt. During the preheat process, you should also keep a close eye on the cowling and paint even if you have directed the hot air away from the cowling, towards the engine. If the engine compartment becomes too warm, this can also cause paint or fiberglass damage. If you notice any signs of damage, shut the preheater down immediately. By periodically laying an ungloved hand on the cowl, you can determine the relative temperature of the cowl and paint. If it is too hot to touch, then it is probably becoming too warm.

You should also be sure the preheater ducting is not pushed against anything inside the cowling as you insert it. Quite often, spark plug wires, spark plugs, primer lines, and fuel injection lines are all exposed and could be subject to damage if the heating duct is improperly inserted into the cowling.

Finally, don't forget to remove the preheater from the engine cowling and move it away from the plane before you attempt to start the engine. This sounds like a step you could take for granted, but every year planes are damaged due to prop strikes of objects near the propeller as the plane starts. In the rush to get the plane's engine started and the cabin heat running, it is quite possible to forget this seemingly obvious step.

As you are preflighting and preheating the plane, you should assure yourself that your plane has had the oil cooler winterized according to manufacturer's recommendations. Most general-aviation piston aircraft have an oil cooler, or radiator, installed in the engine compartment in such a way that air will flow through it and dissipate heat from the oil. In winter, excessively cold air flowing through the oil cooler can prevent the oil from reaching proper operating temperatures. As a result, the oil remains too thick throughout the flight, and excessive engine wear can take place. Many aircraft manufacturers recommend the installation of a plate that blocks all or part of the oil cooler from the incoming airflow. As a result, the oil reaches the correct operating temperature and flows through the engine properly. To determine if this is necessary for your aircraft, review the operations manual and contact your local aircraft mechanic, or contact the manufacturer directly. The installation, if necessary, usually occurs when outside air temperatures become lower than a recommended value.

After you start the engine, you should let the oil warm to normal operating temperatures before takeoff. However, be careful not to let the engine become too warm. When air temperatures are cold, the oil temperature gauge can be misleading. It is possible for the oil temperature gauge to read in the normal range while hot spots form within the engine due to a lack of airflow while on the ground. This overheating can cause engine damage, and this should be avoided. If your plane is equipped with them, cylinder-head temperature gauges can be a more accurate method for monitoring the engine temperature prior to takeoff.

Before leaving this section, we should also discuss a topic somewhat related to preheating—the use of heating devices in the cowl when the plane is hangared. To help prevent engines from cooling down too much as the plane sits in the hangar, some pilots place heating devices inside the cowling. These can range from devices specifically designed to warm the engine and oil, to light bulbs that are slid into the cowling and left on to produce heat. Properly used, these devices can help keep the oil from becoming viscous and thick from the cold, and the devices can reduce the time after start-up that the engine is without proper oil pressure.

Improperly used, these heating devices can be a hazard. Be sure the device is installed and operated correctly. When the heating device generates an open source of heat within the cowling, such as a light bulb, have it positioned in such a way that the heat it generates will not cause damage to the cowling or engine components. If the heat output is too high, it is possible that the heating device could start a fire, so be careful if you are using a makeshift heating device or one not designed for aircraft engines. Finally, be sure to remove the heating device, if not permanently attached to the engine, prior to starting the engine.

Engine start

Starting the engine during the winter can be straightforward if the proper techniques are used for the engine in your aircraft. In this section we will briefly cover some points of interest on this topic. As with other areas, it is not possible to touch on all conceivable situations. The intent is to offer items for consideration as you start your plane's engine.

Most engines require a generous amount of fuel to prime them for start on cold, winter days. If your plane is equipped with a manual primer, it might take five to seven, or even more, pumps on the primer to supply sufficient fuel to the cylinders to cause them to fire. The reason behind priming the engine is that fuel is injected directly into the cylinders, which then vaporizes and aids in the combustion process as the engine starts. The colder the engine, fuel, and air are, the less vaporization of fuel that takes place. As a result, more primer fuel must be pumped into the engine's cylinders to begin the firing process.

For most fuel-injected engines, priming can be accomplished by performing the normal prestarting steps of setting the mixture rich, cracking the throttle, and turning the boost pump on for a few seconds to pump fuel into the engine's cylinders.

Like the manual primer process, this injects fuel into the cylinders and helps begin the combustion process as you start the engine. On very cold days, it might be necessary to let the boost pump run slightly longer.

The amount of prime necessary to start an aircraft engine will vary based on the temperature and engine. As you gain experience with a particular plane, you will be able to anticipate the correct amount of prime. If too much primer is used, it is possible to flood the engine, which will make it more difficult to start. On very cold days this can also frost the spark plugs with fuel, again making it a greater challenge to start. In severe cases, it might be necessary to move the plane to a heated hangar, or preheat the engine, to get the plane to start.

As you attempt to start the plane, do not allow the starter to crank the engine over for extended periods of time. On cold days your battery has less power available to supply to the starter motor, and it will wear down more quickly. Excessive starter cranking can also damage the starter itself. As the starter turns the engine over, it warms. Excessive heat buildup can damage the internal components of the starter, such as the brushes or armature. Extended periods of cranking during the start can also cause unnecessary engine wear. The oil is thick when it is cold, and it provides less wear protection for the moving engine parts. The colder the engine is, or the longer the parts within it move without sufficient oil pressure, the greater the wear that might take place. Use relatively short cranking periods as you start the engine. This procedure will help reduce wear on the battery, starter, and engine.

Finally, once you do start the engine, keep the RPMs initially low until the oil pressure reaches proper levels. This will also help reduce unnecessary engine wear. I have seen pilots use throttle settings that are much too high immediately after the engine starts. This might be under the belief that the cold engine is less likely to stop while turning high RPMs. I cringe each time this takes place, knowing the amount of wear that the engine is experiencing. If your engine does have a tendency to run, and then stop, try the following. As you start to add power to taxi, there is a technique you can try if your plane is equipped with a manual primer. After you prime the engine, leave the primer pump out. As the engine begins to die, slowly push the primer pump in, providing slightly more fuel to the engine. Do not push the pump in too quickly. It is possible to flood the engine, which can aggravate the situation. I have used the technique successfully for a number of years to keep the engine from stalling after starting. It does allow me to use a reduced power setting after starting the engine, which helps reduce engine wear. Be sure to lock the primer prior to takeoff.

Cabin heater/defrosters

Cabin heaters and defrosters for many single and multiengine, general-aviation aircraft tend to be inadequate to the task of heating the aircraft cabin or removing condensation from the windscreen. For most of these planes, the heat is generated by channeling outside air into the engine compartment and over the engine's muffler. The theory is that the hot muffler will warm the air, which is then routed to the aircraft cabin for heating and defrosting. Reality is slightly different. Most planes I have flown have a number of problems when heat is needed. In many cases the heat distribution is poor, resulting in the occupants of the front of the plane being warm, while those in the rear of the plane need additional clothing just to stay warm. During other flights, I have not been able to warm the plane's cabin at all, even with the heat on maximum. On one occasion I ended up putting crumpled newspaper pages around my feet during the flight to help reduce the effect of cold air flowing in from the outside and freezing my feet.

The defrosters on these planes are equally inadequate for the task. When the outside temperature drops, and people inside the plane are exhaling warm, moist air, the result is that windows fog over very quickly. While taxiing, it becomes necessary to frequently rub the windows to be able to see through them. On those flights where the temperature is freezing, this moisture quickly freezes on the windows. Heaters and defrosters might become more effective after the plane is in the air. Most are not equipped with blower motors, and insufficient air flow takes place when the plane is on the ground.

The point of this section is to make you aware of the problems associated with attempting to keep the windows clear and the plane warm during the winter. Be prepared to dress warmly if it is cold outside, and it might also be a good idea to bring along a soft cloth for wiping the windows as they fog over. Each plane can be different, so it is a good idea to be prepared.

Frost/snow on aircraft

Frost and snow are facts of life when flying in a winter climate. Any plane can accumulate frost and snow as it sits out on the ramp. Whether it is exposed to the elements overnight, or for only a few minutes, the accumulation of snow, ice, or frost on your plane can have adverse effects on the plane's flight characteristics. In this section we will review how snow, ice, and frost affect the plane and why. We will relate these effects to the potential hazards they present for planes during takeoff, and we'll cover some of the preventive measures you can take.

Figure 9-2 illustrates how a plane can look with a layer of ice and snow on the wings. Anyone who has seen a thin layer of frost or ice on the wings, tail surfaces, or other areas of a plane knows that these do not add a dangerous amount of weight to the plane. In some cases only a few pounds of frost or snow can thoroughly cover a plane. What is the danger that this thin layer of ice, snow, or frost represents, and why should it be taken so seriously?

Fig. 9-2. *Frost layer.*

In previous chapters we have discussed the airfoil and lift. As you are aware, the air flowing over the wing generates lift, allowing the plane to fly. When the airflow becomes turbulent, such as at or above the critical angle of attack, lift is reduced and the wing cannot generate sufficient lift to keep the airplane flying. Figure 9-3 shows how a layer

Fig. 9-3. *Airflow disruption due to frost.*

of snow, ice, or frost can disrupt airflow over the wing. In this illustration, frost on the inner portion of the wing (arrow) is disrupting the flow of air around the wing root, increasing turbulence, and reducing or preventing the wing's ability to produce lift. As a result, a seemingly insignificant layer of frost can keep the plane from being able to lift off. If the entire surface of the wing is covered with frost, even greater amounts of turbulence are generated. This can prevent any lift at all from being produced.

Every year, pilots ignore the accumulation of ice, snow, or frost on their planes and run off the end of the runway as they attempt to take off. Other pilots manage to get the plane airborne, only to stall and settle back to the ground as they attempt to leave ground effect. This is true not only of general-aviation pilots, but also air-carrier pilots. In some cases, lack of knowledge on the part of the pilots is the reason. In others cases pilots are under the pressure of time or schedules and attempt to fly without properly removing the accumulation first. Whatever the reason, these are avoidable accidents if pilots understand the need to remove the ice, frost, or snow and take the necessary steps.

Depending on the amount of accumulation, you have several options for proper removal. If the accumulation is frost, it is quite possible that it can be rubbed off the plane with a gloved hand. I have done this when fairly heavy frosts have covered the upper surface of the wings and fuselage. For light accumulations, direct exposure to sunlight might generate sufficient warmth to melt the frost.

Many types of snow can also be brushed off, depending on how dense or frozen it is. Do not, as some pilots have, assume that your forward speed during the takeoff run will blow the snow off. There is at least one crash attributed to snow on the horizontal stabilizer that did not blow off as the plane began its takeoff roll. It disrupted the airflow and prevented the up elevator input from raising the nose of the plane, which ran off the end of the runway. The pilots had apparently assumed the snow would be blown off the elevators as they rolled down the runway, which did not take place in this situation.

Ice can sometimes be knocked off with gentle pounding, depending on how thick it is. Be sure not to damage the surface of the plane by striking it too hard. However, if the frost, snow, or ice are too heavy to be removed via the manual methods, it might be necessary to resort to stronger measures. These can include moving the plane to a heated hangar until the accumulation melts, or applying one of the various de-icing fluids that are available from many FBOs. Whatever the correct method is for the situation you are in, use it. Attempting to save time by skipping the removal process could result in the need to spend a great deal of time filling out insurance and FAA accident report forms.

Tires/wheelpants

Tires and wheelpants present an interesting consideration during winter flying that many pilots miss entirely. Ice and snow can cause problems for airplanes that have wheelpants installed, if the pilot does not know what to look for.

When I first began to fly, I noticed that many of the planes at the airport suddenly had their wheelpants removed as winter approached. The reason, the flight school operator explained, was that during the winter, water, ice, and snow can accumulate inside the wheelpants as the plane rolls along a wet or snow-covered surface. With sufficient buildup of the snow and ice, the tires can freeze in place, preventing them from turning. The potential effect of this could range from the plane being unable to taxi, to the wheels being unable to turn as the plane lands. Serious damage to the plane could be one possible outcome. To reduce this as a potential problem, the FBO I was then flying with removed the wheelpants each winter and installed them again in the spring.

If you are flying a plane with wheelpants, and have reasons not to remove them, be sure to inspect them for snow or ice buildups as part of the preflight routine. As you taxi, do your best to avoid areas of ice, snow, or water to help prevent an accumulation inside a wheelpant. If you do find a buildup of snow or ice during the preflight, remove it prior to attempting to taxi.

This same freezing risk due to snow or ice also applies to aircraft brakes. With sufficient buildup, this can cause the brakes to freeze in place. The result could range from a complete loss of braking capability, to the wheels being locked in place. Brake inspection should be part of your standard preflight inspection, but during the winter also look for accumulations of snow or ice that might freeze them in place.

The final topic we will review in the preflight section is how severe cold can affect the tires of the plane. The weight of the aircraft on the tire causes the bottom to be flattened, even with normal air pressure in the tire. If your tire pressure is low, this flat surface can become even larger. In severe, prolonged exposure to the cold, the rubber the tire is made from can freeze somewhat, resulting in the flat spot remaining flat as the tire rotates. Depending on the size of the flat spot and the rotation speed of the tires, this can cause strong vibrations as you taxi or attempt to take off. To remove the flat spot from the tire, you can attempt to warm the tire in a heated hangar until it thaws sufficiently, or you can add air pressure to the tire to bring it up to normal levels recommended by the manufacturer. Some mechanics have recommended temporarily slightly overinflating the tire to remove the flat spot, but be sure this is allowed by the tire manufacturer before you attempt it.

TAXIING

Ice and snow on the ramp, taxiways, and runways cause the same control problems for the plane when you taxi as they do for the car you drive. Steering and braking can be nonexistent as you taxi over slippery patches. In this section, we are going to review some of the hazards associated with taxiing when the ground is covered with ice or snow and what you should do to try and avoid problems.

Let's begin with a situation that doesn't happen frequently but can present an interesting challenge when conditions are right: your plane becomes frozen to the ramp. I have seen it happen when the sun warms the packed snow or ice a plane is sitting on, and the plane sinks slightly into it. After the sun stops heating the ground, it cools and refreezes, now with a firm grip on the tires. I have also seen this take place when wet snow or freezing rain are falling, and as the plane sits on the ramp, the accumulation freezes the aircraft's tires to the surface. Whatever the cause, when this situation occurs, your plane is now part of the ramp. Pilots generally first become aware of this predicament as they apply power in an attempt to taxi. Invariably, the normal power application does not free the plane. The pilot, somewhat confused, thinks, "I'm sure I removed all the tie-down ropes and chocks," and applies more power. In some cases, this addition of slightly more power will cause the plane to free itself, and the plane and pilot taxi away.

However, there are those situations where the plane is firmly part of the ramp, and more thrust will not free it. This is true of not only general-aviation aircraft, but also airliners. I vividly recall an early April day when a blizzard struck southern Wisconsin. I was at the local airport early that morning as they attempted to move a DC-9 back from the gate. With the snow and freezing rain blowing horizontally across the ramp, the towing tug spun its tires as it futilely pushed against the plane. After a short conference among the ground crew and pilots, they disconnected the tug and the plane started its engines. The pilots applied reverse thrust and attempted to back the plane out. With engines roaring and thrust reversers engaged, large chunks of snow were thrown up and the plane shook, but refused to move. After several attempts at this, the engines were shut down and a different approach was taken. The de-icing fluid used on the plane's wings and empennage was sprayed around the tires, trying to melt the ice and snow that anchored it. This finally freed the tires from the ice. The tug was able to push the plane back and it departed. However, if a large airliner can become this firmly anchored, smaller, general-aviation aircraft are also susceptible.

If you find yourself in a similar situation, the primary outcome you want to avoid is damage to the aircraft or landing gear. High-power settings and unnecessary roughness with tugs or other equipment can stress or damage the landing gear. This is even more true for aircraft equipped with retractable gear, where side loads can damage the retraction and locking mechanisms. You do have several options open to you, though, assuming that use of reasonably high-power settings, or use of a tug, do not free the plane.

First, you can try carefully chipping the snow or ice away from the tires. Be careful to avoid damaging the tires as you do this. Quite often this is sufficient to free the plane. Second, I have seen the preheater used to blow hot air around the tires, melting the ice or snow. Again, be careful not to damage the tires by heating them too much. Remember, you only want to melt the ice, not the tires. Finally, I have seen de-icing fluid used to help free the plane. In this situation, make sure the fluid will not cause damage to the tires.

As you taxi your plane along the ramp or taxiway, you should do your best to avoid snow, ice, or puddles of water whenever possible. As we discussed in the preflight section, this material can freeze on the brakes or wheelpants if it is thrown up into them. In situations where this is impossible to avoid, use slower taxi speeds. This will not throw the ice, snow, or water as far and will help reduce the amount that is deposited on the brakes or wheelpants. It can also freeze on retractable-gear mechanisms, making it impossible to raise the gear after takeoff, or later lower it for landing.

Ice and snow also present hazards related to steering and braking as you taxi. Just as your car can begin to skid out of control on a slippery patch, the same thing is possible in your plane. Use slower taxi speeds and apply brakes sparingly as you travel over slippery surfaces. Give yourself plenty of room to make your turns, and avoid having the nosewheel skid along sideways as you attempt to turn the plane by using wider-radius turns. This skidding places side loads on the nosewheel that might result in damage. If you start to skid, reduce power and let the plane come to a stop. Do not aggravate the situation by adding power or locking the brakes. Also, as you taxi, give yourself extra distances to brake the plane to a stop. If the plane begins to skid across the ramp as you apply brakes, your only option might be to let it come to a halt without brakes. If you are too close to another plane, building, or edge of a runway or taxiway, you might run out of room before the plane comes to rest, so give yourself as much distance as possible from other objects as you taxi.

When you are taxiing in high-wind conditions, weathervaning might be a real problem for both tricycle and tailwheel planes. Due to the reduced braking and steering action available as a result of snow or ice, what might have been acceptable wind conditions during the summer could get out of control very quickly. Even with brakes and steering, you might not be able to maintain the direction you want while you taxi. Consider the surface, wind strength, and direction you want to move as you prepare to taxi. Use sufficiently slow taxi speeds and anticipate what the wind is going to attempt to do as the plane moves along the ramp or taxiway.

When you perform the runup, be prepared to reduce engine power if the plane begins to skid across the ramp on the slippery surface. It might be necessary to use reduced power settings as you do the mag check and other runup steps to prevent this from happening. Common safety practices and sound judgment can help you avoid taxiing accidents during the winter.

TAKEOFF RUN

All of the considerations we have discussed for taxiing, such as turning and braking, apply for the takeoff run as well. Make sure you are straight on the runway as you begin your takeoff roll, and as soon as enough airspeed has been achieved, use flight controls to maintain directional control. Until this point, use brakes and nosewheel steering minimally. What seem to be minor control inputs can quickly cause the plane to lose directional stability and begin to skid. Use of flight controls, instead, will provide more positive control that is not dependent on tire traction along the runway to keep the plane straight as it moves down the runway.

Soft- and short-field takeoff procedures will change little from those described in previous chapters. If the snow is rutted or uneven, it might have a tendency to pull the plane during the takeoff roll in much the same manner as described in the chapter on soft-field operations, chapter 7.

This presents an additional chance for skidding to take place because the plane might change direction, but momentum carries it along the original heading. Try to use the smoothest portion of the runway; avoid the rough or rutted areas.

During takeoff you will find that the takeoff distance and angle of climb can be substantially different than during the summer. Takeoff distances can be greatly reduced due to the higher-density air that is present during winter. After getting used to the performance that a particular plane exhibits during the summer, it can seem as though it leaps from the runway on a cold, January day. You might also find that the pitch angle can seem greater to maintain an airspeed such as V_X or V_Y. In some cases this can be a startling change in aircraft performance. Again, colder, denser air will allow the plane to climb at a steeper angle than during the summer. Some students who begin learning during warmer months fail to maintain the correct airspeed during climbs in the winter. Rather than flying the correct airspeed, they fly the pitch angle to which they have become accustomed. Be prepared for the different performance characteristics and maintain control of the plane.

One special concern is crosswind situations. As with taxiing, the crosswind can cause the plane to weathervane, and the reduced steering and braking capability might not give you sufficient control to overcome the wind's effects. A strong crosswind can also push the plane sideways, completely off the runway. This effect will be strongest with a direct crosswind. Be sure that you use the correct crosswind-control positioning to help prevent loss of control. If crosswinds are too strong, waiting until they subside might be the best decision. Depending on how little surface traction exists, it might be necessary to cancel a flight even when the crosswind component is well below the demonstrated crosswind component for your plane.

LANDING ROLLOUT

Landings on snow- or ice-covered runways are another unique aspect of winter flying that can be executed more safely with the proper experience and understanding. In chapter 4 we discussed hydroplaning on wet runways and, in many respects, hard-packed snow and ice present many of the same control issues. During takeoff on these surfaces, you are building up speed at a steady rate, with the plane hopefully under control from the beginning of the takeoff run. During landing you are touching down at a high speed, and on a slippery surface the plane is at greater risk of skidding across the surface. To help minimize loss of control during the landing and rollout, you should work to keep the plane tracking straight down the runway from the initial touchdown. Use the smallest control inputs possible. Large steering or brake inputs while you are moving at a high speed can cause the plane to begin skidding down a slippery runway.

Figure 9-4 depicts a plane skidding down an icy runway after landing. There are many ways to induce a skid after touchdown. It is not possible to describe how to avoid or get out of every skid situation, but in this section we will review some basic steps that can be taken to reduce your chances of getting into skids. We will also cover how to get out of skids. Depending on the situation you find yourself in, it might be necessary to modify these techniques, or to use a different set of procedures entirely. In any situation the most important thing you can do is keep thinking and maintain control of the plane. Once you let the situation begin to control you, the chances of regaining control become more difficult and less probable.

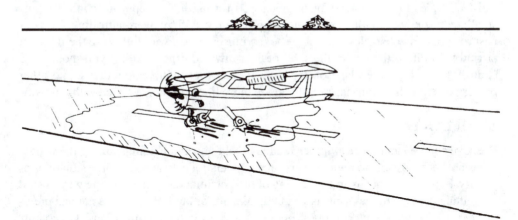

Fig. 9-4. *Skidding airplane.*

The best way to avoid a skid is to be aware that it can happen. As you are coming in to land, know what the runway conditions are prior to touchdown. At tower-controlled airports, the ATIS frequency will normally indicate what braking conditions are or the amount of snow or ice cover on the runway. At uncontrolled fields, it might be necessary to ask on CTAF for that information. If there is no one that can answer, it can be a good idea to make a low pass down the runway to determine what shape it is in. By knowing how poor conditions are, you can plan your approach accordingly.

Keep your airspeeds as low as possible during touchdown in slippery conditions. Just as a car is more difficult to control in a high-speed skid than in one at a slower speed, a plane will be more controllable after touchdown at slower touchdown speeds. This works well for both soft- and short-field landings, which are designed to touch down at the slowest possible airspeed. Make sure you touch down with the plane's longitudinal axis parallel to the runway. If you have a tendency to land with the plane slightly off the runway heading, the plane will have a greater tendency to skid.

This becomes even more true for crosswind situations, where improper crosswind techniques can aggravate the situation. Poor runway alignment, or drift across the runway, can turn a landing into a white-knuckled ride down the runway. Once you touch down, keep your steering inputs as small as possible to reduce your chance of overcontrolling the plane. Use brakes sparingly to slow the plane. If you have sufficient runway, let the plane slow without the use of brakes. Braking too heavily can also lead to a skid.

Try to avoid ice- and snow-covered portions of the runway as you roll out. Braking action will be better on the dry, clear portions of the runway. As with takeoffs, ruts, potholes, and rough areas on the snow or ice can cause the airplane to swerve out of control with no warning. By knowing runway conditions prior to landing, you can help reduce the chances of a skid developing. If you should enter a skid, use the same control techniques that were previously discussed for taxiing. Keep your control inputs small, reduce power to idle, and do not aggravate the skid by overcontrolling. Do your best to avoid any obstacles that might be in your path. Runway lights are the frequent target of aircraft as they leave the sides of the runway during a skid. If you should lose control, try to keep the plane on the runway. Runway lights, snow drifts, and other planes present collision hazards that you are more likely to hit if you leave the runway.

SUMMARY

We have covered a number of topics related to takeoffs and landings during the winter. It has not been possible to review every area, or technique, to avoid trouble situations as you fly. Experience and an understanding of the potential dangers go a long way toward preventing damage to the plane as you taxi, take off, or land. If you have not had previous experience with winter flying, be conservative as you take off or land. Give yourself and your plane an extra buffer of distance in all maneuvering. Snow drifts along the ramp, taxiways, and runways can reduce your wingtips to nothing more than shattered fiberglass or twisted metal. Low-wing planes are even more susceptible to this type of damage than high-wing models. Some runways seem to shrink in width dramatically as snow is plowed into mounds along their sides during snow removal. When this happens, you have little room for error as you take off or land. Figure 9-5 illustrates just how narrow a runway can become. What had been a sufficiently wide runway during the summer is no longer adequate with large mounds of snow lining the runway. Even when the runway is wide enough, all of the takeoff and landing techniques, such as touching down on the runway centerline, become even more important in these situations.

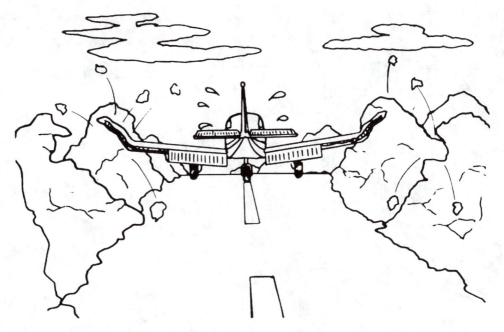

Fig. 9-5. *Narrow runway.*

We have not even touched on some of the more specialized considerations, such as the use of snow skis on the landing gear. Before you attempt to fly with this type of aircraft configuration, be sure to get instruction from a qualified flight instructor. It can substantially change the performance characteristics of the plane and operating techniques.

Know your aircraft's heating system and use it properly. Breathing alone can quickly frost over the windshield, and some aircraft do not have defrosters that are capable of removing heavy deposits. I have had situations during which it was necessary to open the pilot window and look out of it from time to time during the approach to see where I was in relation to the runway. If your windows do frost over, do not use sharp objects to scrape the frost off. That might scratch or gouge the plexiglass. I have used my gloves or hands to rub the frost off. Sometimes this works well. Other times it does not. Make sure the defroster and cabin heat are on at maximum settings to help avoid this potential problem.

To conclude, it you are not sure about a given situation as you fly in the winter, ask a flight instructor who can help you find the answer to your question. Flying in the winter can present you with smooth air and beautiful vistas, but be prepared as you operate during the winter, and you will be rewarded with another positive flying experience.

10
Abort procedures

ABORT PROCEDURES ARE SOME OF THE MOST IMPORTANT PROCEDURES that you should know, yet they are the least understood and practiced. Whether problems arise during takeoff or landing, you should be automatic in your response to the situation, smoothly completing each step of the abort procedures. Unfortunately, many pilots never give abort procedures a second thought once they complete their private-pilot rating. As a result, when a situation requiring them to abort a takeoff or landing arises, they fail to take necessary action or wait until it is too late to realize the situation even exists.

Previous chapters have made reference to the classic landing abort, the go-around. In this chapter we will review a number of topics related to aborted takeoffs and landings. This will include reasons that an abort might be necessary and the general abort procedures. We will also review potential problem areas that you should avoid during the abort situations. When you complete the chapter, you should have an understanding of when an abort might be necessary and the general abort procedure steps. Please keep in mind that it is not possible to cover every situation or aircraft in this review, and it might be necessary for you to adjust procedures, depending on the actual situation. Let's begin with the takeoff.

TAKEOFF ABORTS

Of the two abort situations, takeoff aborts are probably less practiced than landing aborts. Many pilots do not even consider what they would need to do if the engine quit during the takeoff roll, and they do not look for any of the warning signs of impending engine problems as they advance the throttle for takeoff. If the problems are recognized at an early stage, it is much easier to safely abort the takeoff and stop the plane on the runway. In this section we will discuss these topics in greater detail.

Causes

Takeoff aborts are generally caused by a situation that will prevent the plane from safely achieving takeoff speed in the available runway. As you are traveling down the runway, you should be constantly monitoring the available runway ahead of you, airspeed, rate of acceleration, and engine instruments. Before you begin the takeoff run, you should also have set a point on the runway beyond which you will abort the takeoff if you have not become airborne. This takeoff-abort point should allow adequate distance to safely brake the airplane to a halt.

A number of factors can precipitate the need for an aborted takeoff. Some of these the pilot can determine prior to ever starting the airplane, while others are true emergency situations that must be dealt with in an appropriate manner. In chapter 6 we reviewed the use of takeoff- and landing-distance performance charts. High-density altitude is a very real potential cause for insufficient runway being available for your takeoff. Lower engine horsepower, reduced thrust, and reduced lift will result in the need for additional runway during takeoff. If you review the performance charts for the plane you fly, you will find that takeoff distances increase dramatically as the density altitude increases.

Yet every year, pilots run off the end of the runway as they attempt to take off on runways that are too short for the density altitude they are operating at. If the pilots had accurately determined the takeoff distance ahead of time, these accidents could have been avoided. At some point during the takeoff run, the pilots should have realized they were not going to get the plane off the ground. They should have aborted takeoff, but they failed to do so. There can be several reasons this takes place. First, the pilots had not considered the need to abort the takeoff, and they never considered the option as the takeoff progressed. Instead, they continued down the runway, their minds set on completing the takeoff. Second, the pilots did not monitor the rate of acceleration or airspeed during the takeoff run. As a result, they were not aware of the aircraft's poor performance and missed the cues that would have helped them understand that they needed to abort the takeoff. Third, the pilots might not have recognized that they were not going to leave the ground before they ran out of runway, and they waited too long to abort the takeoff. In this situation, abort procedures are executed, but indecision on the part of the pilot prevents them from performing them early enough in the takeoff to be effective. Finally, even though the pilots determine they will not be able to complete the takeoff, they do not attempt to abort. In this case, pilots often freeze and do not remain in control of the situation. At this point the plane often ends up running off the end of the runway before it comes to a halt. If the pilots are lucky in this situation, the only damage is to the plane.

There are other factors that can result in the need to abort a takeoff. As you advance the throttle, you should scan oil pressure, fuel pressure, and other engine instruments. Some pilots lock the brakes, advance the throttle to full power, scan the gauges and, if all gauges are in the green, release the brakes. This is a fairly common practice for turbine-powered aircraft, but it seems to be less popular for piston-engine planes. Other pilots advance the throttle, then scan the gauges as the plane rolls down the runway. Both methods can give you important information concerning the health of the engine and the option to abort the takeoff well before it becomes impossible to do so. By scanning the engine gauges early in the takeoff run, the aircraft's speed will be lower, and less stopping distance will be necessary during the abort.

Unfortunately, some pilots merely push the throttle forward and assume the engine is performing correctly. The reliability of today's aircraft engines is extremely high and, given the number of general-aviation planes, relatively few pilots have found it necessary to deal with partial or complete engine failure during takeoff. For this reason, pilots become complacent and do not plan what steps they will take if the engine should have problems. However, engine problems do occasionally develop during the takeoff run and, if you become aware of them early enough, it is possible to avoid more serious situations such as running out of runway or attempting to land when the engine quits just after lift-off.

I like to scan the engine gauges after the throttle is advanced to full power, then again prior to the plane leaving the ground. In this manner I am aware of any engine problems and can react accordingly. You should verify that oil pressure, fuel pressure, engine RPMs, manifold pressure, and other engine instruments are in their green ranges. If they are not, abort the takeoff immediately. As the throttle is advanced and the plane is rolling down the runway, you should also listen to the engine. There are occasions where the gauges might all look good, but the engine does not sound right. Fouled spark plugs, damaged spark plug leads, and damaged magnetos are just a few of the items that can cause a reduction in engine power during the takeoff. It is much easier to make an early recovery from an aborted takeoff than to attempt to nurse a malfunctioning engine to keep the plane in the air until you are able to get around the pattern to land.

One year while I was attending the Oshkosh EAA Fly-in, there was a flight of P-51 Mustangs taking off. A nearby radio broadcast one of the Mustang's pilots calling out he had low fuel pressure and was aborting the takeoff. The pilot smoothly braked the plane to a stop, and turned off the runway. This is an example of a pilot performing the necessary scan during the takeoff run, finding a problem, and successfully resolving the situation by aborting the takeoff. This might very well have prevented the situation from becoming worse if the plane had made it into the air, where a true emergency could have developed. I also know two other pilots who experienced engine failures immediately after takeoff. In both situations, the pilots were able to land.

In one case, the pilot was injured and required hospitalization for recovery. The point is that engine problems during takeoff do happen. You should be mentally prepared to abort the takeoff if necessary, and you should scan engine gauges during takeoff for signs of trouble.

Before moving on to abort procedures, we should briefly discuss other situations that might require you to abort the takeoff. In previous chapters we discussed looking for other aircraft traffic or objects on the runway as you execute the takeoff. During the takeoff you should scan intersecting runways and the runway ahead of you. If another plane presents a potential hazard, or if a vehicle or animal are located on the runway ahead of you, it might be necessary to abort the takeoff to avoid the potential danger of impacting another plane during takeoff. Again, this is part of the mental preparation that should be part of each takeoff you perform.

Takeoff abort procedures

There are general procedures you can apply if you find it necessary to abort the takeoff run. The plane you fly might have a different set of procedures. If this is the case, use those recommended by the manufacturer. This point becomes even more true if you are flying a multiengine aircraft, the procedures for which are not reviewed in this book. First and foremost, maintain control of the plane. In many accidents, pilots become so focused on staring at a gauge, or the item related to the emergency, that they forget to control the plane's direction of travel. In these cases the plane then frequently leaves the runway, the pilot not even realizing the direction the plane is heading until it is too late.

As you steer the plane down the runway centerline, reduce power to idle, then begin maximum braking. Keep in mind that maximum braking is braking that is the hardest that can be done without the tires locking up and skidding. As you apply maximum braking, it might be necessary to input aft elevator control to reduce any nose-over tendency the heavy braking might induce. Conventional-gear planes will be more susceptible to this than tricycle-gear planes. If you have flaps extended for takeoff, retract them. This will help reduce lift and increase braking.

Depending on the situation, it might be wise to perform additional steps once the plane is under control. If there is danger of a fire from impact, leaking fuel, or leaving the runway, pull the mixture to idle cutoff. In addition, turn the fuel selector valve to off. These steps will cut off fuel to the engine and help reduce the potential for fire as a result of fuel leaking into the engine compartment. The engine will also stop running and the propeller will stop turning. This will reduce the chances for prop strike damage to the propeller and engine. If you are flying at a controlled field, you should radio to the tower to let them know you have experienced problems and might be blocking the runway. At an uncontrolled field, use CTAF to make others aware of your situation. If possible, as you coast along to a stop, try to turn the plane onto a taxiway to keep the runway clear.

Once the plane has halted, it is a good idea to exit it and inspect the plane for leaking fuel or damage. I have seen pilots not shut the engine down and taxi back to the ramp, but whether you should do this will depend on the type of emergency you have experienced. If oil pressure is low, shutting the engine down might be the best course of action to avoid damaging it due to oil starvation. Low fuel pressure might not be as damaging and, if the engine continues to run, it might be possible to get back to the ramp before shutting the engine down.

I have had only one instructor have me perform takeoff abort practice over the years. You might have a difficult time finding a flight school that will perform takeoff abort practice because of how potentially hard it might be on the plane. In many cases the instructor will pull the throttle to idle just before the plane reaches flying speed. At this point, the student should perform the abort steps previously discussed. The heavy braking practiced during the maneuver can be hard on the plane, though, and many flight schools wish to avoid that amount of wear on the brakes and tires. However, there are those flight schools around that will teach takeoff abort procedures, so it would be a good idea to find one that is willing to work with you.

LANDING ABORT (GO-AROUND)

As a student, your primary flight instructor should have had you practice aborted landings many times. The purpose of practicing the go-around is to learn to be able to safely abort a landing when, for some reason, it is no longer safe to attempt to land on the runway. This situation can arise for many reasons. Other traffic taxiing onto the runway, traffic taking off or landing on intersecting runways, landing too far down the runway to stop safely, and severely bounced landings are only a few of the many reasons that can necessitate the need for a go-around.

The problem is that many pilots do not practice this maneuver after they complete their private rating, and they no longer have a flight instructor making them do it. Like the aborted takeoff, the pilot begins to forget that a go-around presents an option that might be required during any landing. Finally, when a situation arises that calls for a go-around, they continue in an attempt to salvage the landing. In some cases luck intervenes and they are able to avoid a serious problem, while at other times they are not. In this section we will review some of the potential causes for a go-around and the general procedures you should apply during one. It is not possible to review every situation or go-around procedures for every aircraft. If you are out of practice or unfamiliar with the procedures for the plane you fly, find a qualified flight instructor to work with you.

Causes

While any number of reasons might make it necessary to execute a go-around, there are several common situations that affect pilots on a regular basis. We will discuss these more common scenarios in this section. This will help increase your awareness of what you should be looking for during the approach to landing.

The most common reason I have aborted a landing is due to another aircraft taxiing onto the runway as I am landing. This seems to happen all too frequently in busy pattern situations, where the pilot of the offending airplane is attempting to take off between the plane that has just landed and the plane on short final. We discussed this particular situation in chapter 3, which covered traffic patterns. I doubt that any pilot would purposely force another plane to go around, but in some cases the other pilot's estimated window of opportunity for takeoff is larger than the one that actually exists. In my own experience, I watch as the plane taxis onto the runway and immediately wonder if its rate of acceleration far exceeds the normal aircraft's capability. At the same time, I prepare myself for the inevitable go-around I know I will soon be executing. Depending on the set of circumstances, I might execute the go-around immediately or continue in on final until I am certain the other plane will not be able to get out of the way before I reach the runway. The timing of the go-around depends on the distance between you and the other plane and how quickly you are closing on it.

There are several things to be aware of as you execute a go-around in this situation. First, once you determine that a go-around is necessary, do not hesitate to commit to it. Second, as you execute the go-around, keep the plane that is taking off in sight. It is quite possible for the plane that is taking off to climb into your plane if you do not remain aware of the other plane's position relative to yours. The other pilot will probably not be able to see you and might not be aware you are there. For this reason it is up to you to maintain adequate separation. Third, watch for other planes in the pattern. Pilots tend to space themselves within the flow of the pattern. The go-around you are executing will disrupt the flow of traffic and the spacing that other planes have set up in the pattern. For this reason, as you turn to reenter the pattern you might find yourself in very close proximity to other aircraft. Scan the airspace as you work your way back into the pattern and be certain to maintain adequate spacing from other aircraft. Don't focus your attention on the irritation you are likely to feel at having to abort the landing. Don't forget about flying the plane; always remain in control of it.

Another reason you might need to execute a go-around is if there is traffic taking off or landing on an intersecting runway. We have previously discussed the need to scan intersecting runways for other traffic during the approach. When this situation exists, there is a potential for a collision at the intersection of the runways. If you do find that other traffic poses a danger, go around. Do not assume the other traffic sees you or is aware of you. I have found that, when possible, communicating with the other traffic can help reduce the need for a go-around. By finding out what their intentions are and letting them know of yours, you can often avoid the need to execute a go-around. But if there is any question as to the safety of a landing, abort the landing and set up for another approach.

Animals, or other objects, that mistakenly end up on the runway you are landing on can also cause problems during landing. Due to their wide open areas of land, airports frequently offer a haven for many animals. During some seasons they might also migrate across the airport and runways. I have seen deer and fox on airport runways, and running into one of them after touchdown could cause serious damage to the plane and its occupants. Once you turn onto final, scan the length of the runway and the area around it periodically to ensure that there are no animals, airport vehicles, or other objects on or uncomfortably near the runway. Animals might panic and run into the plane if you continue the landing. Deer do serious damage to cars every year, and it is very possible they will also bolt toward your plane after you touch down.

Landing too far down the runway to safely get the plane down and stopped is another popular reason for aborting an attempted landing. All too often, pilots are determined to get the plane down on the runway, no matter how far down the runway they are going to touch down. In this situation the pilot is normally attempting to force the landing, not considering the inability of the plane to stop in the remaining runway. Reasons for this narrow vision during the landing include fixation on the landing to the point of ignoring where the plane is going to touch down, lack of understanding of the runway length requirements for the plane, panic, and poor judgment. No matter what the reason, the plane normally lands so far down the runway that it runs off the end or is damaged during the attempt to stop it before it leaves the runway.

When you are landing, pick your point of intended touchdown and the point beyond which you will execute a go-around. This becomes critical in short-field landings, where any miscalculation on the touchdown point can result in a potentially serious situation. Whether you are floating down the runway due to high airspeed, or you just missed the intended touchdown point entirely, once you have reached the go-around point, do not hesitate to execute the procedure. Smart pilots are the ones who understand that a compromised situation exists. They use the go-around as an option to allow them to set up properly the next time. Pilots that attempt to force the landing, or believe that their piloting skills will come into question if they execute a go-around, are those who end up filling out accident reports.

Just as landing too long is cause for a go-around, so is landing short of the runway. In this situation, for one reason or another, the plane will touch down short of the runway. In some cases, pilots attempt to stretch the glide by pulling back on the elevator. This can result in the plane stalling and dropping in very hard, or worse. In other situations, the pilot ends up landing short of the runway. If you are going to land short, execute a go-around if it is not a salvageable landing. Again, once you make the decision to go around, do not hesitate to perform it. You will probably be flying at a low airspeed, and any delay might result in the plane touching down before you bring in full power.

One of the other very frequent situations that can result in a go-around is the bounced landing. Every pilot has experienced that occasional landing resulting in the plane coming back up off the runway and seeming to hang in the air at an uncomfortable altitude, angle of attack, and airspeed. This situation presents an excellent opportunity to execute a go-around and try the approach again. I have watched student pilots freeze as the plane floats upward, airspeed decaying and a bone-jarring drop imminent. When pilots feel this unsure of the outcome of the landing, it is time to add power and fly away. When pilots do not go around in a badly bounced landing, but instead choose to ride the plane down, they are setting themselves up for a potentially bad situation.

Some pilots have developed a stigma related to a go-around after a bounced landing. They feel that somehow they are less of a pilot if they are unable to salvage a landing and need to go around. Nothing could be further from the truth. Some of the best pilots I have met have no problems at all with executing a go-around when it is necessary. Even in conversation with others, they will explain that if a landing is not working out as intended, execute a go-around. To them this is the mark of a competent pilot. So remember, if you bounce a landing, keep the go-around as an option.

Any badly executed approach is sufficient reason to execute a go-around. If the airspeed is too low or too high or the glideslope is off, abort the landing attempt and set up again. This can also include remembering at the last second that you forgot to lower the gear. You might also find that as you get down close to the runway, the crosswind is stronger than you or the plane are able to correct for. It is not possible to cover every reason that you might need to execute a go-around, but whenever you do not feel comfortable with an approach, keep the go-around as an option.

Go-around procedures

The plane you fly might have procedures that are somewhat different from the general procedures we are going to review in this section. If that is true, use the recommended procedures for your plane. However, many of the concepts we discuss will likely hold true for almost every general-aviation aircraft.

To begin, once you make the decision to execute a go-around, do not hesitate. The first step is to add full power immediately. If you have carburetor heat on, turn it off after power has been applied. This will ensure that the engine is able to develop full power. As you apply power, make sure you watch manifold pressure, engine RPMs and, if equipped, turbocharging. You do not want to overstress the engine and cause damage as you execute the go-around. After you have established full power, there are several steps you will want to take simultaneously. First, change the attitude of the plane to slow or stop the rate of descent. Then allow it to accelerate to a safe climb speed, such as V_X or V_Y. Depending on how you have trimmed the plane, it might have a very strong tendency to pitch up and turn to the left. The pitch-up tendency is due to the increase in power and the nose-up trim you have added during the approach. It could require a significant amount of forward pressure on the control yoke or stick to overcome this pitch-up moment. The left-turning tendency is due to P-factor and the other left-turning forces that will be acting on the plane. As you overcome the nose-up tendency and maintain the correct airspeed, be sure to keep the plane flying in a coordinated manner to avoid having it stall/spin as a result of uncoordinated use of the controls. Figure 10-1 displays how the plane can pitch up severely after power is applied. When you are low to the ground, this is a very bad situation to get into.

Some student pilots have commented that they hesitate to overpower the pitch-up tendency because they feel that they might damage the plane. This is not true. By allowing the plane to pitch up too high, the airspeed will bleed off and the plane might stall and impact the ground. Maintain control of the plane and make it fly the correct airspeed. After full power and the correct airspeed and attitude have been established, retrim the airplane to reduce the need for forward pressure on the control yoke. This

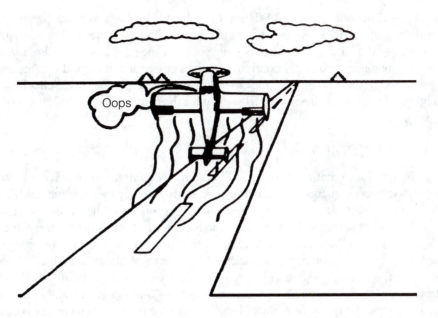

The hammerhead stall is an interesting maneuver. . . but not on a go-around!

Fig. 10-1. *Aborted landing pitch up.*

will make it easier to hold the correct climb airspeed and will reduce the tendency for the plane to pitch up. Initially, this retrim can be a rough trim setting. The point of the retrim is not to set it to hold the airspeed exactly, but to reduce the amount of pressure necessary to reduce the pitch-up tendency.

After you have achieved sufficient airspeed, gradually begin to reduce flaps if you have deployed them during the approach. This should be done incrementally to avoid an excessive loss of lift, which could cause the plane to settle to the ground. Again, as you retract the flaps, maintain the correct airspeed and allow it to accelerate after you remove each notch of flaps. If your plane is equipped with retractable gear, wait until after the flaps are retracted and you have established a positive rate of climb before raising them.

For two reasons, flaps should be fully retracted prior to raising the gear. First, if the plane should end up settling to the runway as you execute the go-around, you will want the gear down to prevent a gear-up impact. This can happen if power is applied too late, the airspeed is too slow, an incorrect nose-up attitude is held, or flap retraction causes a reduction in lift that results in the plane losing altitude. Second, flaps generate more drag than landing gear. By cleaning up flaps before the landing gear, you will achieve a positive rate of climb more quickly and find it easier to hold the desired airspeed. You should also have completed a rough retrim of the plane prior to retracting the landing gear.

As you climb out, you should fly on the right side of the runway to watch for any other departing traffic, as we previously discussed. Once you are back in the pattern, be sure to use the landing checklist as you set up for the next landing. It is easy to assume that the plane is correctly configured after a go-around, then find that the prop is not set correctly, the gear is up, or the cowl flaps are not set. By using the checklist, you will be certain that the plane is properly set up for the next landing ("On Landings: Part II," p. 7).

PILOT INDECISION

The technique for executing a go-around is relatively straightforward. By performing them in the correct sequence, you can transition from a landing to a positive rate of climb with minimum altitude loss. Most pilots get into problems by hesitating when a go-around becomes necessary. There might be many reasons for this hesitation: a desire to salvage the landing, lack of understanding of the situation, panic, and poor judgment. As a result, the go-around might be executed too late to be of any usefulness. Commit to aborting the landing as soon as you realize it is necessary. Figures 10-2 and 10-3 illustrate how waiting too long before committing to a go-around can result in excessive altitude loss and the inability to clear obstacles near the runway. You can see that if the pilot had executed the go-around when he or she became aware of the need, the plane could have cleared the tree at the end of the runway ("On Landings: Part II," p. 7).

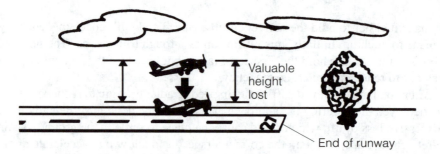

Valuable height lost

End of runway

Fig. 10-2. *Delayed go-around effect on altitude.*

SUMMARY

In this chapter we have reviewed abort procedures for both takeoff and landings. There are two important points you should remember from this chapter: follow the procedures that are recommended for aborting a takeoff or landing and, once it becomes necessary to abort, do not hesitate to take action. You should also regularly practice aborts for both takeoffs and landings to maintain your proficiency in both procedures.

The procedures we reviewed for aborted takeoffs and landings are general in nature. If your aircraft's manufacturer recommends different procedures, be sure to use

Fig. 10-3. *Effects of indecision during go-around.*

those. If you are at all rusty or unsure of the procedures, take a qualified flight instructor along to fly with you until you are proficient. While the purpose behind practicing the procedures is to be able to execute them safely, if they are not properly performed they pose the opportunity for problems to arise. Be sure to keep safety as the primary consideration as you practice.

11
Emergency-landing procedures

EMERGENCY-LANDING PROCEDURES ARE ONE OF MY FAVORITE TOPICS TO teach student pilots. When flying with students, I frequently pull the throttle and simulate an engine out, then I have them execute emergency-landing procedures. In this chapter we are going to cover a number of topics related to emergency landings. These topics will include engine-out warning signs, rugged-terrain landings, retractable-gear problems, and other related areas.

When you have completed this chapter, you should have a stronger understanding of the correct procedures to use for various engine-out landings. If you maintain control of the plane, you and your passengers should be able to walk away from almost any forced landing in a single-engine airplane. I speak from experience on this particular topic. Less than a year ago I experienced an engine-out situation and made a forced landing. I used the techniques I was taught as a student pilot and, as an instructor, have been teaching to my students. Later in the chapter I will relate the full story, but it is quite possible to walk away from emergency landings if you keep your head and maintain control of the plane. Let's begin by discussing the scans you should do as you are flying.

PILOT SCAN

As you fly, there are several scans you should perform. Pilots are taught to scan the instruments as part of their private-pilot training. In the previous chapter, we discussed scanning engine gauges during the takeoff roll. However, you should also scan them frequently during your flight. This scan should include not only flight instruments, but also fuel gauges, oil pressure and temperature, cylinder-head and exhaust-gas temperatures, and any other gauges related to the engine.

Many pilots get out of the habit of doing the scan, especially after they have been flying for some time on a cross-country flight. Other pilots forget the gauges after completing the runup, assuming that if everything was in the green then, it will remain that way. However, every year pilots end up making forced landings as a result of fuel starvation or the engine quitting for a host of other reasons. In the next section, we will discuss some of the warning signs of imminent engine problems. By scanning the instruments and gauges frequently, it is possible to recognize a potential engine problem early enough to avert a forced landing.

You should get into the habit of doing the scan every few minutes as you fly. At first you will need to make a conscious effort to do it, but eventually it will become a habit. I have even gotten into the habit of doing the scan while flying inverted in the Pitts Special I fly.

There is a second scan that you should also do as you are flying, but in this case it is outside the plane. Years ago I read an article in an aviation magazine about an instructor who pulled the power on a student and asked where he should make an emergency landing. The student replied he was flying over rugged terrain and there was no place to safely land. The instructor then asked why the student had flown into that area if there was no suitable site to land. That article left an impression on me. I began looking around for places to land as I flew, trying to remain conscious of the terrain I was flying over and how suitable it was for an emergency landing. In time, this became another scanning habit I developed—looking at the terrain I was flying over as a matter of habit. Now, as I scan for other aircraft, I also look at the terrain around me for suitable landing sites.

I firmly believe that pilots should habitually scan for potential landing areas as they fly. After practicing this for a while, it becomes an almost subconscious activity, much like the engine-gauges scan. In the Midwest, where I have done much of my flying, there are so many corn and bean fields that it is hard not to have a suitable field for an emergency landing. Over cities this becomes more of a problem, but even then, roads often present adequate emergency landing strips if you are thinking about how to use them to your advantage.

However, in other areas of the country, the terrain can become much less hospitable. You might be flying over heavily wooded areas, marsh, rough terrain, or other equally rugged ground. If you should lose an engine when flying over these types of land, a forced landing becomes dependent on quickly finding the best area available and correctly performing forced-landing procedures. Time wasted in an extended scan looking for a potential landing area could result in sufficient altitude loss and might prevent you from reaching the best area available. Periodically scanning the areas you are flying over will help reduce the time and altitude you would otherwise lose before making your forced-landing decision.

As you scan, look for the areas that present the best emergency-landing sites. This will, of course, vary a great deal, depending on the type of terrain you are flying over. Generally, the flattest, most wide-open area should be your first choice. However, you might find that this is not an available option. In wooded areas you might be forced to land on top of the trees. In rougher terrain it might be necessary to land on the side of a hill or along the side of a river. Overall, you should be looking for a landing area that will result in the least damage to the plane and passengers.

Mountain flying is a science of its own. Many pilots who intend to fly in mountainous areas take specialized training from flight schools specializing in mountain training. In those courses you learn some of the specialized mountain flying techniques that can help you deal with forced landings. If you intend to do any flying in mountains, I strongly recommend that you find a reputable flight school that specializes in teaching pilots to fly in those areas.

By scanning during the flight, you become aware of the terrain you are flying over. You might find that it is a good idea to slightly alter your flight path to avoid a particularly rugged area in favor of more hospitable areas. These small deviations in your course might add a small amount of time to the flight, but they will greatly increase your flight's overall safety.

WARNING SIGNS

Unless your airplane experiences a sudden, catastrophic failure of a major engine component, such as the crankshaft or a connecting rod, there are normally warning signs that the engine is having problems. Earlier in this chapter we discussed the need for periodically scanning engine instruments. If the engine is having problems, the scan can make you aware of them, in some cases with sufficient time to land before the problem becomes a true emergency.

Over the years, I have experienced one complete engine failure and a second situation where the engine gave a great deal of notice that a problem was imminent. In the first situation, a sliver of brass worked its way through the fuel system and became lodged between the needle and seat in the carburetor, resulting in the engine being completely flooded. There were no warning signs prior to when the engine quit. Oil pressure and temperature were both normal until the engine stopped. I realized there was a problem when I advanced the throttle to level off after a descent, only to find the engine was not producing any power.

During the second situation, I was flying a load of skydivers up to jump altitude. While performing my normal scan, I noticed the oil temperature was beginning to rise to above-normal temperatures. It was a warm day, and I had made a number of jump runs that day, so initially I attributed it to the engine just being warm. I lowered the nose to let the engine cool somewhat. As I continued to scan, I watched the oil pressure begin to drop, not uncommon if the oil temperature is high. However, the temperature was not abnormally high, so this began to point to other problems. Finally, the constant-speed propeller began to fluctuate in RPMs, the engine began running rough, and I was unable to maintain altitude. The propeller governor uses oil pressure to control the pitch of the blades, another indication that there were oil system problems.

When flying skydivers, I circle within a mile or so of the airport as the plane climbs. The reason for this is that if there are engine problems, the jumpers can exit the plane and normally make it to the drop zone under canopy. The other reason is that the plane can be flown back to the runway. (In my opinion, that's the more important reason, since the pilot stays with the plane.) At this point I let the jumpers know about the problems and set up to let them out of the plane over the airport. Once they were away from the plane, I pulled the power to idle to reduce any possible damage to the engine and glided into a landing. A mechanic was called to look the plane over. As it turned out, the engine had recently been replaced and the oil dipstick from the previous engine had been installed in the new engine. This dipstick indicated the oil level was several quarts higher than it actually was. The mechanic explained that even though the stick read that the engine was full of oil, it was actually several quarts low, resulting in the problems I experienced. After the oil was properly filled, no further problems were found.

As you scan the engine gauges, you should be looking for abnormal indications. High or low readings on the oil pressure or temperature, fuel pressure, cylinder head temperatures, or other gauges are the first signs that a problem exists. By correlating the readings, you can often determine how serious the problem is. Readings of low oil pressure, coupled with high oil temperature, are evidence of oil system problems. However, if just low oil pressure is indicated, without the rise in oil temperature, you might only have a faulty oil pressure gauge or sensor. In the first situation, the engine might quit in the very near future, while in the second you might be able to get to an airport and land. In either case you should land at the nearest airport while scanning for suitable landing sites in case the engine quits while en route. Oil is the lifeblood of the engine, and by monitoring those gauges, you can get a solid indication of its overall health.

Other gauges can also relate the engine's fitness to you as you scan them. If engine RPMs fluctuate on a constant-speed propeller, you might have oil system problems or a propeller governor problem. Low fuel pressure might indicate a blocked or leaking fuel line or a problem with the fuel pump. If your plane is equipped with a boost pump, try turning it on to see if the fuel pressure improves. Low manifold pressure on a constant-speed propeller, or low RPMs with a fixed-pitch prop, can indicate carburetor ice. In this case you should apply carburetor heat and see if readings return to the normal level. Remember, carburetor ice can take place in temperatures up to 70°F or higher, especially on humid days, so do not discount that as the cause of a power loss on warm summer days.

A faulty vacuum pump can cause low vacuum pressure, resulting in inaccurate readings from vacuum-driven instruments such as the artificial horizon and directional gyro. During the runup, the low vacuum pressure should be indicated by the vacuum pressure gauge, but it is also a good idea to scan this gauge during flight. On one occasion, while flying VFR on top, I began to feel that the artificial horizon was not correct. This is a common feeling as part of vertigo, so I concentrated on the gauge and did the instrument descent through the clouds. Once beneath them, it became readily apparent that the artificial horizon was indeed wrong. Cross checking with the vacuum gauge showed marginal vacuum pressure.

High exhaust-gas temperatures might indicate that you have leaned the engine too much. In this case, *detonation* (the uncontrolled, explosive burning of fuel in the cylinder) might take place, resulting in higher exhaust-gas temperatures and loss of power (Fig. 11-1). The danger of excessive leaning and detonation is that the pistons, connecting rods, and crankshaft might be damaged as a result. Most aircraft engines should only be leaned to maximum settings when the engine is producing at or below 75% maximum power. Be sure to consult the operations manual for the aircraft you fly to determine proper leaning procedures for your plane. Prolonged detonation can cause engine damage that might require you to make a forced landing. By monitoring the exhaust-gas temperature gauge, you can help reduce the chances of this taking place. Monitoring can also be very useful as you climb or descend, when it might be necessary to adjust the mixture as the air density changes.

Normal combustion

Normal burning

Fig. 11-1. *Detonation.*

Explosion

Detonation

In this section we have reviewed a number of different warning signs the engine can give you that problems might exist. It is not possible to cover every situation, and we have only reviewed some of the basic guidelines. It should be readily apparent that in order to see these signs, you must be looking for them. This is the reason scanning instruments periodically is so important. Quite frequently, the engine will let you know something is wrong far enough in advance to avert an actual emergency. By scanning, you can take advantage of this and save yourself a lot of adrenaline.

ENGINE LOSS ON TAKEOFF

Engine loss during takeoff is, in most situations, the worst possible time to experience engine failure. You will be at a low altitude with a relatively low airspeed; you will have little time to react, and few options for landing areas are available to you. For this reason alone, runups are extremely important. A runup will help determine if the engine is running properly and will give you warning signs of potential problems. There are, however, those situations where you have done everything correctly, and the engine fails without warning shortly after lift-off. In this section we will review some of the FAA-recommended procedures that you should take if you lose your engine during takeoff.

When talking to pilots, there seems to be two ways they believe they should handle this type of engine-out situation. One reaction is to attempt return to the airport by executing a tight, 180° turn to get back to the runway. The reasoning behind this is they feel it is possible to get the plane around quickly enough to safely get to the airport and land. This group of pilots is of the opinion that this is safer than attempting an off-airport landing.

The second reaction to loss of engine is to attempt an off-airport landing somewhere ahead of the airplane. The rationale behind this method is it is unsafe to attempt a steep turn so low to the ground, and the chances for success are better by finding the best place ahead of the plane to land. I have talked to pilots in both camps, some exhibiting tremendous conviction that their method of choice is the appropriate one.

The FAA has published a document, "Impossible Turn," document number FAA-P-8740, that I would highly recommend to any pilot. In this section we will discuss the contents of that document, which clearly shows the dangers of attempting to return to the airport after an engine loss during takeoff. A number of studies have been done to determine the best course of action the average pilot should take when faced with an engine loss at takeoff. In short, in most situations the safest reaction to an engine loss is not to attempt to make it back to the airport for a landing, but to find a suitable site within a 120° scan around the plane.

Studies have shown that it typically takes a pilot approximately four seconds to react to loss of engine after takeoff. During that four seconds, the pilot is attempting to understand what has happened. Of course, the reaction time will vary among pilots, but this is a typical reaction time. In the course of that four-second interval, several things begin to happen. First, the plane will start to lose airspeed immediately. It is at a relatively high angle of attack during the climb and, as a result, it will begin to lose airspeed rapidly. Without thrust, the plane will also begin to lose altitude. How rapidly will depend on a number of factors, but you and the plane will definitely begin to start down immediately.

After you have determined that you no longer have an engine providing thrust, there are several actions you need to take. The first is to adjust pitch to achieve the best glide airspeed. If you do not adjust the pitch, the plane will slow and, rather quickly, stall. Low to the ground, without an engine, is not the place to practice stall recovery techniques. Once you have achieved the correct airspeed, you are faced with a choice: attempt a turn back to the airport or make an off-airport landing somewhere ahead of the plane's current direction of flight. The facts in this case are startling in how great the odds are against your successfully completing the return to the airport.

We will review the example listed in the "Impossible Turn" document. Let's assume you are at 300-feet altitude when the engine stops. If you are determined to return to the airport, you must decide how tight the turn should be. Steeper turns will result in a smaller turn radius, but greater altitude loss. A standard rate turn will result in less altitude loss, but a larger radius. Figure 11-2 illustrates a comparison of two different turn radiuses. Example 1 is a standard-rate turn made at a 70-knot airspeed, which results in a turn radius of 2240 feet. In addition to the 180° turn, it will also be necessary to make an additional 45° turn to head back toward the runway. The time required to complete this turn are 4 seconds for the initial reaction delay, 60 seconds for the standard-rate turn, and 15 more seconds for the last 45° to the runway, for a total of 79 seconds. The text further states that if the plane is descending at 1000 feet per minute, during the 79 seconds it takes you to complete this maneuver, the plane will lose a total of 1316 feet of altitude. Since you started at only 300 feet in this example, it quickly becomes clear you will not make it back to the runway.

What if we increase the rate of the bank and tighten the turn up? To begin, in previous chapters we discussed that as the angle of bank increases, so does the stall speed of the aircraft. To avoid stalling during the steeper turn, you will need to maintain an airspeed higher than the 70 knots we used in the first example. To achieve this higher airspeed, you will need to lower the nose, further increasing the rate of descent. Example 2 in Fig. 11-2 shows that the radius decreases to 560 feet in a 45° banked turn and requires only 10° additional turn to make it back to the runway. Before you become too optimistic, let's review the altitude loss numbers and the time to complete the turn. The 45° turn will result in a turn that is approximately four times that of a standard-rate turn and require 15 seconds to complete. Add to this the 4 seconds for the initial reaction and another second for the last 10° of turn, and the total time turns out to be 20 seconds. Even using the 1000-foot-per-minute descent in the first example, this results in an altitude loss of 333 feet. Since we began this example at 300 feet, the results are still unacceptable. Even if you think just getting close to the runway is better than an off-airport landing in this example, couple coming up short of the runway with a very high rate of descent and the plane being in a steep bank close to the ground. It is much safer to touch down in a controlled landing attitude in a field or open area than close to the airport in a steep bank. The likely outcome is the plane striking a wingtip first, then cartwheeling out of control, or performing some other equally unacceptable maneuver before coming to a stop.

The information just presented should make you stop and seriously reconsider the potential you have to successfully make the return to the airport from a low-altitude loss. In addition to the difficult turn and getting the plane into the correct landing atti-

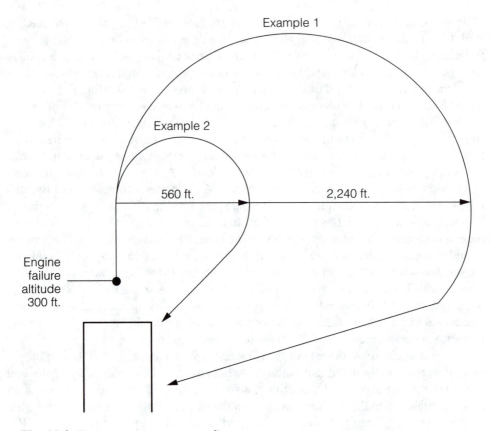

Fig. 11-2. *Return to runway turn radiuses.*

tude, there are also downwind and crosswind landing factors to take into consideration. The turn radiuses previously reviewed were for no-wind situations. Turning with the wind will further increase the radius of the turn. What direction the wind was from will likely be the last thing you consider as you begin the turn back to the airport. A downwind landing will also result in a higher ground speed at touchdown, increasing the potential for damage to the aircraft and occupants.

At this point you are probably wondering, "What is a safe altitude to attempt a return to the airport?" This particular document from the FAA indicates that at least a 600-feet altitude is necessary before you can even begin to have a reasonable chance of success. To make the airport, you will need to establish a 9°-per-second turn, achieve the correct glide angle immediately, and add 10 knots to the speed used for a standard approach. Even from this altitude, the document cautions that this be done only when confronted with unacceptable landing options in all other directions. The chances for success, even in this situation, will depend a great deal on the characteristics of the plane and how well the pilot reacts. I know that the Pitts Special I fly has a very steep rate of descent without power, and to attempt a 180° turn from 600 feet would be difficult. Other airplanes that have very good glide characteristics might offer higher chances of success. To make a valid judgment, you will need to know the plane you are flying and how it reacts to this type of situation.

If you are faced with an engine loss with insufficient altitude to make the return to the airport, what are the steps you should take? Some documents in the past have indicated that your only option is to land straight ahead and take what the situation gives you. This seems like a very limited approach to dealing with an engine out, when it is quite possible that a suitable landing site might be available if your course were altered slightly to the left or right. The "Impossible Turn" document suggests that a realistic method for dealing with the engine out is to scan the surface ahead of you 60° to the left and right of the initial course for the best landing area available. Figure 11-3 illustrates the options that might be available to you within the 120° of arc this scan encompasses. As you can see, by altering course to the right, the plane in this example has a suitable landing area available to it.

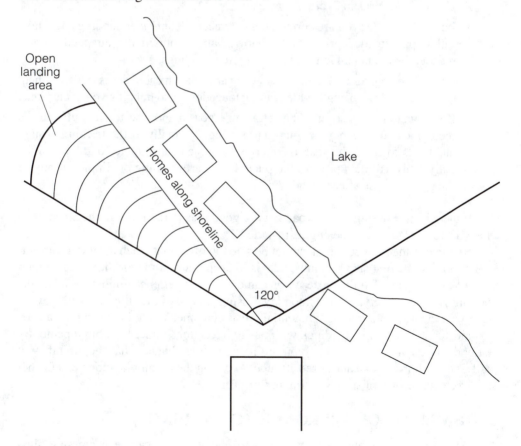

Fig. 11-3. *Takeoff engine failure landing scan.*

Keep in mind that the lower your altitude when the engine quits, the less time you will have to make the scan and complete the turn. Below the 300-feet altitude, it might not be possible to safely maneuver the plane 60° in one direction or another, and you might need to limit the scan area to less than 120°. It does no good to select a landing area if the plane has insufficient altitude to turn and reach it.

When making a forced landing, you should follow the emergency procedures for your plane. These are normally outlined in the operations manual. However, there are a number of general steps that should normally be taken. These include:

1. Set the pitch of the plane to achieve the best glide speed. This will give you the most gliding distance for the altitude you are at.

2. Scan for a potential landing area within the 120° arc previously discussed.

3. Pull the mixture to idle cutoff, and turn off the fuel and magnetos. These actions will help reduce the chances of a fire.

4. Use shallow turns to avoid obstacles.

5. Lower flaps after you are certain you can reach the intended landing site. Have full flaps extended prior to touchdown, and maintain the airspeed recommended by the manufacturer for emergency landings.

6. Prior to landing, turn off the master switch and unlatch cabin doors. This will help prevent them from being jammed during touchdown and trapping you in the plane.

7. If the plane you are flying is a retractable-gear plane, use the manufacturer's recommend emergency-landing-gear setting. Normally, if the terrain is rough and likely to cause the gear to collapse, keep the gear retracted and land on the belly. This will help prevent the plane from cartwheeling or flipping if any of the gear collapse after touchdown.

If there is time during the approach, it is a good idea to run through the emergency checklist to attempt to get the engine restarted.

Before closing this section, it cannot be said too often that when an emergency situation occurs, the first step you must take is to keep your mind working and maintain control of the plane. Don't freeze up or panic if you experience engine loss. Let the training you have received come into play. Losing control of the situation will only reduce your chances of avoiding a potentially bad outcome. While luck can play a role, it is much better to couple that luck with positive reactions rather than blind panic. By maintaining control of the plane, you are reducing the chances that the aircraft will stall, spin, or impact at an undesired attitude. Avoiding these complications will further increase your chance of success ("Impossible Turn," p. 2–5).

ENGINE LOSS ON APPROACH TO LANDING

Engine loss during landing can be very easy to deal with, or it can be as pressure filled as loss of engine after takeoff. If your pattern has been flown in a manner that allows you to glide to the runway after an engine loss, then it remains very easy to make it to the runway and execute the landing. However, if the pattern you have flown has put you far enough away from the runway that you are not able to glide to it, then a real problem exists. In this section we will discuss how you should deal with the loss of your engine during the approach and how you can avoid setting yourself up to be short of the runway if you should lose an engine.

Loss of the engine while landing is near and dear to my heart. Not long ago I was completing a sightseeing flight with a young family and was returning to the airport for a landing. I had eased the power back to the bottom of the green arc and was descending as I approached the airport. I applied power to level off, only to find that the engine did not wish to cooperate in the power department. I proceeded to check fuel, switch fuel tanks, and turn on carb heat, all to no avail. While doing this I trimmed the plane for best glide and tried to fly a straight line to the airport, hoping I could make it.

The propeller was windmilling, so the passengers were not aware that a problem existed. As I continued toward the airport, I began to review the terrain around me for additional potential landing areas. There were a couple of suitable cornfields close to the airport and a nice, wide, four-lane highway that could also be used. I continued in, finally deciding that I could make the airport, but none of the runways. There was enough grassy land on the airport that it made the best choice for the landing. Just prior to touchdown I let the passengers know we had an engine problem and would be landing in the grass. I set up for a soft-field landing and let the tower know of my problems. As the plane slowed on short final, the prop shuddered to a stop and I touched down in the classic, nose-high, soft-field landing attitude. We rolled to a stop and exited the plane, none the worse for wear. As I previously mentioned, a piece of brass had become lodged between the needle and seat in the carburetor and flooded the engine to the point that it quit running.

In this particular case, the tower had cleared me to land, but I had not set up for a pattern yet due to how far out we were from the airport. If we had been out slightly farther when the engine quit, I would have ended up using either the cornfields or the highway as a landing area, but on that day luck was with me and the engine quit at a most opportune time. I have always tended to fly a tight pattern in relation to the runway. During my initial training as a private pilot, one of my instructors explained the wisdom of this type of close pattern in the event of an engine out, and I have followed that line of thinking whenever the traffic and tower allow me to. I also tend to make my approaches power off, feeling that if I am planning to glide to the runway to begin with, loss of the engine will present no additional challenge at that point.

While it is not always feasible to fly patterns in this manner, doing so as often as possible reduces the chances that I will not be able to glide to the airport if I should lose the engine. There are times when the tower or traffic in the pattern make it impossible to fly a pattern close to the runway, but it is something that should be done whenever possible. Even more so than before the engine loss, I work very hard at staying close to the runway during pattern work and try to make my students do the same. Occasionally, as I instruct, I will pull the power on downwind to bring the lesson home to the student. There is not much worse than watching the runway slide up on the windscreen, knowing you will not be able to glide to it as a result of flying a wide pattern. This helps students understand that if the pattern had been flown slightly tighter, they could have made it to the runway by gliding.

If you should lose the engine and are too far from the airport to make it there, you should immediately set the plane's attitude for the best glide speed and begin looking for a suitable landing spot. Depending on your altitude, you might have a fair amount of time to find a good landing spot and make an approach. You should follow the same steps for the emergency landing that were described for engine loss at takeoff, unless your aircraft's manufacturer recommends a different set of procedures. These steps typically include:

1. Airspeed: Set to best glide

2. Mixture: Idle cutoff

3. Fuel selector: Off

4. Ignition switch: Off

5. Wing flaps: As required

6. Master switch: Off

7. Doors: Unlatched prior to touchdown

8. Touchdown: Soft-field attitude

9. Brakes: Apply heavily

However, if you have flown a pattern that is close enough to the runway, you should find it possible to glide to the runway and land on it. You should still perform the steps just documented, but you will make a normal runway landing. If the propeller is windmilling, be prepared for it stopping as the airspeed slows. It can make the plane shudder as it comes to a halt. However, do not be startled by the shaking that might take place. Remember to maintain control of the plane until it comes to a complete rest.

I can vouch for the fact that the emergency landing training you were given while obtaining your private pilot rating works. During my engine-out landing, I used the techniques I was taught by my primary flight instructor, and I have taught these methods to my students.

Dangers of dragging it in

In previous chapters we have discussed what can happen if you drag the airplane in on landing. Figure 11-4 illustrates what can happen if you lose the engine while flying a low, slow, high-power approach. As you can see, if the engine suddenly loses power, the glideslope changes drastically and the plane will not make it to the runway. This is another point in favor of flying approaches that allow you to glide to the runway if you should lose the engine. It might not always be possible to fly this more favorable approach, but it should be done whenever possible.

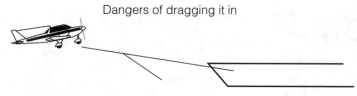

Dangers of dragging it in

Fig. 11-4. *Engine loss on final.*

Some will say that it is easier to put the plane on the numbers if you drag it in on a shallow approach, chopping the power just as you cross the end of the runway. Hitting the numbers, or any other designated spot on the runway, can be done as consistently, and with more safety, by flying a more normal glideslope during the approach and using the approach and touchdown techniques that we have been over many times in this book. By using this method, you are relying on your piloting skills, planning ability, and judgment to hit the intended touchdown point. You are not relying on the engine's power. In the end, flying approaches in this manner will result in more consistent landings and will improve your skills.

RUGGED-TERRAIN LANDINGS

Making emergency landings in rugged terrain can require that you alter the previously discussed emergency-landing techniques. Standard touchdown and landing procedures work well for flat, open areas, but they present a potential hazard when making a landing in a heavily wooded area, over rough or rolling ground, or when making a water landing. In the following sections, we will review the steps you should take when faced with each of these particular emergency situations.

Trees

If you should lose your engine while flying over heavily wooded terrain, it might be impossible to find an open area available for the emergency landing. There are, however, recommended procedures you can take to minimize damage during touchdown. The plane should be flown just above the trees, followed by a full stall. The full stall reduces the plane's forward speed as much as possible and allows the plane to drop in a more vertical line onto the tops of the trees. The stall height above the trees should be as low as possible to reduce to a minimum the vertical descent speed that the plane will be dropping at. FAA personnel recommend that the approach and flare be flown in much the same manner as any approach to a runway, with the plane being flared just above the treetops and a full-stall landing made on top of them.

The emergency-landing procedures should include:

1. Airspeed: Set to best glide
2. Mixture: Idle cutoff
3. Fuel selector: Off

4. Ignition switch: Off

5. Wing flaps: Full

6. Master switch: Off

7. Doors: Unlatched prior to touchdown

8. Touchdown: Full stall

The procedures are almost identical to the normal forced-landing steps previously covered. You should use full flap extension to reduce forward speed as much as possible, and touch down in a nose-high attitude.

Obviously there are no guarantees when making a landing onto treetops, but use of these procedures will help reduce the chances for injury. A friend of mine with the FAA told me about a pilot who lost an engine over heavily wooded terrain and performed the treetop landing technique just described, making a full-stall landing onto the treetops. The plane was totaled, but the pilot ended up with only minor injuries. When the emergency took place, he followed the procedures he had been taught, and they worked as they were supposed to.

Rough ground

When landing on more rugged terrain, it can be difficult to find a suitable landing site. Hilly, rocky, or other types of land can present a real hazard during a forced landing. As previously mentioned, if it is possible, avoid these areas as you fly. While this might not always be possible, it can reduce your risks. If you do fly over these types of areas, fly as high as practical. In addition to giving you a greater view of the surrounding area, this also gives you more gliding distance and a potentially larger number of landing sites to choose from.

It is not possible to cover all types of emergency landings in rough terrain, but there are some points you should take into consideration. First, follow the engine-out landing checklist for your airplane. This might be similar to the general procedures we have already covered, but use the correct procedures for your plane. Second, find the best landing site immediately after the engine quits, and set up for it. Delay in picking a landing site might result in not being able to reach a suitable area due to loss of altitude. Make your initial scan wide enough. Students frequently scan too narrow an area during engine-out practice, missing potentially good landing sites. Third, be sure to touch down as slowly as possible. As we just discussed in the tree-landing section, a full-stall landing might be the best approach you can fly. Fourth, avoid as many obstacles on the ground as you can. While this might seem obvious, in the heat of the moment it can be forgotten. If at all possible, try to keep the cabin area from taking any head-on collisions. If your plane has retractable gear, it might be wiser to land with the gear up when landing on rugged terrain, for reasons we have already covered. Fifth, in any emergency-landing situation, try to avoid a downwind landing whenever possible. This will increase your ground speed, even though your airspeed might be at a minimum. It might not always be possible to land into the wind, but try to avoid downwind landings.

When you are setting up for an emergency approach into an open farm field, check to see if the rows of corn, or other crops, are straight. If you see rows that are not, these are likely on some type of hilly terrain, which causes them to not look straight from the air. Figure 11-5 represents the difference between how these fields will look from the air. A nice, long field with rolling hills might not prove as suitable a landing site as a shorter, level one. If you are forced to land on hilly terrain, including runways with slopes, the FAA recommends landing uphill if wind and obstacles allow it (Fig. 11-6, "On Landings: Part II," p. 5).

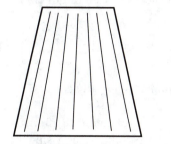

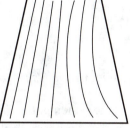

Fig. 11-5. *Rolling hills in a field.*

Flat field rows

Field with hill on right

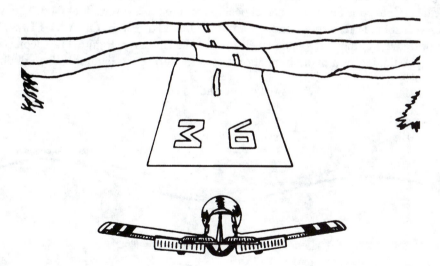

Fig. 11-6. *Uphill landing.*

Each emergency landing situation into rugged terrain is unique and will require that you adapt to it. If you maintain control of the plane, land as slowly as possible, find the best site available to touch down in, and keep thinking throughout the entire process, you will increase your chances of a successful outcome to the emergency.

Water landings

Water landings present some unique aspects to an emergency landing. You will want to use the same emergency-landing procedures we have previously reviewed, but take some additional factors into consideration. Part 1 of the *Airman's Information Manual* (AIM) contains detailed procedures that you should use when making a water landing, or ditching. This section will not cover all the information contained within that section of the AIM, but it will give a summary of the highlights you should be familiar with. It is highly recommended that you read through this entire section of the AIM for complete details.

To begin, there are three primary factors that affect a successful aircraft ditching. These include:

- Sea conditions and wind.
- Type of aircraft.
- Skill and technique of the pilot (*Airman's Information Manual*, 1993).

If you are forced to make a water landing, the first step you should take after setting the plane up for the best glide speed is to review the surface of the water. Using the proper heading during touchdown will help reduce damage to the plane and the chance of injury to you and your passengers. The swells, or rolling motion of the waves, need to be taken into consideration when choosing the direction you want to land. The AIM states you should "remember one axiom—AVOID THE FACE OF A SWELL." The face of a swell is the side of the swell toward the observer. Figure 11-7 depicts the proper positioning of a plane in relation to swells during the landing. When ditching

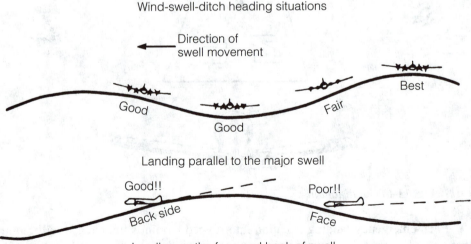

Fig. 11-7. *Proper swell positioning.*

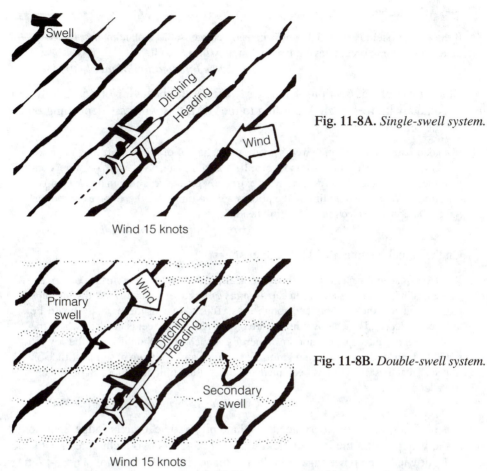

Fig. 11-8A. *Single-swell system.*

Fig. 11-8B. *Double-swell system.*

parallel to the swell, there is little difference whether the touchdown is on the top of the crest or in the trough. The preferred landing is on the top or back side of a swell whenever possible. Your landing heading should have the most headwind component possible while still flying parallel to the swell (*Airman's Information Manual*, 1993).

Figure 11-8A depicts a single-swell system, while Fig. 11-8B represents a double-swell system, both with winds of less than 15 knots. The majority of sea conditions involve two or more swell systems running in different directions. Here the AIM states:

> One of the most difficult situations occurs when two swell systems are at right angles. For example, if one system is eight feet high, and the other three feet, plan to land parallel to the primary system, and on the down swell of the secondary system. If both systems are of equal height, a compromise may be advisable—select an intermediate heading at 45° down swell to both systems. When landing down on a secondary swell, attempt to touch down on the back side, not the face of the swell.

> (*Airman's Information Manual*, 1994)

211

The AIM further states:

If the swell system is formidable, it is considered advisable, in landplanes, to accept more crosswind in order to avoid landing directly into a swell.

(Airman's Information Manual, 1994)

Other areas of this section of the AIM discuss the behavior of an aircraft as it makes contact with the water. According to the AIM, this will vary depending on the state of the sea:

If landed parallel to a single swell system, the behavior of the aircraft may approximate that to be expected on a smooth sea. If landed into a heavy swell, or into a confused sea, the deceleration forces may be extremely great—resulting in breaking up of the aircraft. Within certain limits, the pilot is able to minimize these forces by proper sea evaluation and selection of ditching heading.

(Airman's Information Manual, 1994)

On the actual landing, the AIM further states:

When on final approach the pilot should look ahead and observe the surface of the sea. There might be shadows and whitecaps—signs of large seas. Shadows and whitecaps close together indicate short and rough seas. Touchdown in these areas should be avoided. Select and touch down in any area (only about 500 feet is needed) where the shadows and whitecaps are not so numerous. Touchdown should be at the lowest speed and rate of descent which permit safe handling and optimum nose-up attitude on impact. Once first impact has been made, there is often little the pilot can do to control a landplane.

(Airman's Information Manual, 1994)

It is difficult for even experienced seaplane pilots to accurately judge their height above the water during the landing, and the AIM suggests two techniques for the actual touchdown. When power is available, the plane should be flown in a nose-high attitude approximately 9° to 12° pitch up, and 10% to 20% over stalling speed until the tail makes contact with the water. At this point, power is cut and the aircraft lands. This technique is used by seaplane pilots and helps avoid misjudging the height and dropping the airplane in from a considerable height. If engine power is not available, the AIM indicates you should use a higher-than-normal approach speed down to the flare:

This speed margin will allow the glide to be broken early and more gradually, thereby giving the pilot time and distance to feel for the surface—decreasing the possibility of stalling high or flying into the water.

(Airman's Information Manual, 1994)

The complete text of the water-ditching section of the AIM is several pages in length and is reading material you should review in its entirety. This section of the book is intended to give you a general overview of water landing procedures. By reviewing the complete AIM documentation, you will gain a broader knowledge of the topic. To summarize, while the approach can be flown in a relatively normal manner, the touchdown is flown in such a way as to reduce the chances of damage and injury. In discussions with FAA inspectors, they indicate that if you are flying a fixed-gear plane, there is a strong chance it will flip onto its back during the touchdown. If you are flying a retractable-gear plane, be sure to make the landing with the gear up.

RETRACTABLE-GEAR PROBLEMS

Retractable-gear planes offer speed advantages over those planes equipped with fixed landing gear. By streamlining the plane, drag is reduced, and the plane is normally able to fly at higher cruise speeds. There are those occasions, though, where mechanical, electrical, or hydraulic problems can prevent the gear from extending normally. When these situations come along, it is normally possible to extend the landing gear via other means.

If you find that you are not able to extend the gear once you are in the pattern, the first step you should take is to exit the pattern. This will remove you from the flow of traffic and reduce any tendency to continue the approach without taking the time to adequately review the situation. You should review the emergency-gear-extension procedures for your plane and be familiar with them, but the following are some general procedures you might find helpful.

First, place the gear handle in the "down" position. On some planes there are "emergency" extension procedures that will either release hydraulic pressure, mechanical locks, or other means of holding the gear in the retracted position. Some aircraft are equipped with nitrogen bottles that give the pilot a one-shot means of forcing the gear into the down and locked position via compressed gas. There have been instances, though, where the pilot failed to put the gear handle in the down position before using the nitrogen gas, which prevented it from lowering the gear ("On Landings: Part III," p. 3). Other aircraft might have gear that free-falls into the extended position. I have seen instances were it is necessary for the pilot to take additional measures to force free-fall gear into the locked position. In some cases, pulling the nose of the plane up generates enough G-force to force the gear down and lock it, while in other cases it is necessary to sideslip the plane to use airflow to lock the gear. This information should be in the operations manual for the plane you fly.

The key point to remember is that it is normally possible to get the gear down if the system fails. I was once getting ready to take off at the local tower-controlled airport when a P-51 Mustang experienced problems with its gear extension. The pilot left the pattern and was able to get the gear to free-fall into place. He then did several tower flybys to assure the gear was down and locked. After a successful landing, an inspection found that a hydraulic line had ruptured and no pressure was available to lower the gear through normal means. I have had several instances where the gear indicator light was faulty, and jiggling it, or testing it with another gear's indicator light, proved that the gear was actually working correctly and only the indicators were faulty.

If it becomes necessary to land with gear up, use the emergency procedures we have reviewed for a forced landing. To reduce damage to the propeller and the engine, if possible the propeller should not be turning as you touch down. Landing in grass might be less damaging than on a hard-surface runway, but this will depend on what is available. Properly executed, it is possible to accomplish a gear-up landing with only minor damage to the plane's underside.

In the event that only one of the main or nosewheel landing gear will not extend, you are faced with a choice. You can retract the gear and land on the belly, or you can land on the extended gear and attempt to hold the plane in the correct attitude to keep a wing or the nose from striking the ground while there is sufficient airspeed and control authority. I have read accident reports of pilots who were forced to land with one main gear that would not extend. Landing with only one main gear and the nosegear extended, they were able to hold the wing off the ground until there was not enough aileron control to keep the wing level. At that point the wing is lowered as gently as possible. Some damage to the wingtip normally takes place, but if flown with enough finesse, the damage is minimal. You should avoid letting the wing drop too quickly or at too high an airspeed, as this will increase any tendency for the wingtip to dig in and cause the plane to groundloop or cartwheel.

When faced with landing without the nosegear down and locked, it is often possible to hold the nose off the runway or grass until the plane is at a very slow forward speed. Again, to help prevent damage to the propeller and the engine, the propeller should not be turning. In some cases all that is necessary are minor repairs to the cowling. As with lowering a wingtip to the ground too early, lowering the nose to the ground prematurely might increase the damage to the plane or increase the chances of nosing the plane over. Land in a nose-high position and hold the nose off the ground as long as you have elevator control, then gently lower it to the ground.

There are many gear problems that we have not covered in this section, and many that would be specific only to a particular aircraft. If you think you are experiencing gear problems, use the radio to work with observers in the tower or on the ground prior to landing. They might be able to confirm that the gear is or is not working correctly. If you do experience a true gear problem, use the emergency gear procedures for your plane and maintain control of the plane until it comes to a stop.

FLAT-TIRE LANDING

If you should blow a tire on takeoff, you might find it necessary to land on it if you become airborne just as it goes flat. While it would be better to abort the takeoff prior to lift-off, sometimes the circumstances do not permit this option. In this event you will want to follow some of the procedures we discussed for a main-gear-up or nosewheel-gear-up landing. By holding the flat tire on the main gear or nosewheel off the runway until the plane slows, you can reduce damage to the plane.

Figure 11-9 shows how you can hold a main landing gear off the runway to help prevent the flat tire from contacting the runway until the plane slows. Here the pilot is keeping the wing with the flat tire raised during the approach and flare. Use of aileron can normally hold a main gear off until a relatively slow airspeed, in the same manner

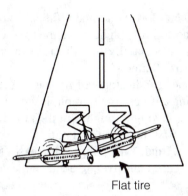

Fig. 11-9. *Flat main landing.*

Flat tire

that we discussed landing with one main gear off the runway during crosswind landings. This will help reduce damage to the wheel rim. Once the plane comes to a stop, you should not attempt to taxi on the flat tire because this might further damage the rim and require that it be replaced. If you have a flat nosewheel, you should also hold the nosegear off the runway with the elevator until the plane slows as much as possible, then lower it gently.

If possible, shifting passengers or baggage to the rear of the plane might help keep the nosewheel off for a longer period of time before you lower it to the runway. When you have a flat main gear, burning fuel from the wing tank over the flat gear and shifting passengers might reduce the speed the wheel must be lowered at. These same principles might also be useful for landing with a main gear or nosewheel that will not extend. Be prepared for any directional changes the plane might attempt to go through due to the drag of the flat tire as it contacts the runway ("On Landings: Part III," p. 3).

DOWNWIND LANDINGS

While not necessarily an emergency landing situation, downwind landings are an option that pilots need to be aware of. For reasons such as approach obstacles or terrain, a pilot might find it necessary to land downwind instead of into the wind. Very tall trees, buildings, or cliffs at one end of a runway might make it difficult or impossible to land into the wind. Before you attempt it, there are several factors that you should be aware of that will affect your approach, flare, touchdown, and rollout during downwind landings.

First, let's examine the major reason for making landings into the wind. Landing into the wind helps to reduce the landing distance necessary to get the plane you are flying down and stopped. This includes the ground-roll portion of the landing and landings where you are clearing an obstacle on final. The reason for this is that when you are flying into the wind, your ground speed is reduced by the same value as the headwind component. Depending on how strong the winds are, this can be a significant drop in your speed relative to the ground. For instance, if the headwind you are flying into during final approach is 20 mph, and you use 70 mph on final, the ground speed of the plane will be 50 mph. If you land downwind under the same conditions, the ground speed will be 90 mph, a difference of 40 mph between the two scenarios. It quickly becomes obvious that it will take substantially less distance to stop with a ground speed of 50 mph at touchdown versus 90 mph.

During downwind landings, your airspeed indicator will not be affected by the wind. In both downwind and upwind landings, if you use 70 mph, it will still indicate that value relative to the mass of air you are moving through. But the tailwind will affect how you should fly your approach. The first thing you will notice is that it will be necessary to tighten up your turn from base to final as the tailwind increases the ground speed of the plane. As with any ground reference maneuver, to achieve the correct turn radius you will need to use a steeper bank, or begin your turn sooner, to come out on the runway centerline. If you do not, it is quite possible you will end up well off the runway's centerline as you complete the turn.

Assuming that you maintain the same relative distances from the runway during the pattern legs, you will find that the plane covers the distance over final much faster than you are used to. This is once again due to the higher ground speed. You might also discover that you will land much farther down the runway than normal if you fly the approach in a standard manner, reducing power and setting flaps at the normal intervals during the approach. The additional ground speed can result in a somewhat flatter relative glideslope as compared to landing into the wind. To compensate for this, you will need to turn to final farther out from the end of the runway, reduce your power earlier, or apply flaps sooner that you normally do. These techniques will help you dissipate unnecessary altitude without the need to dive at the runway to make your landing target.

As you get close to the ground, it will seem as though the plane is moving much faster than normal, compared to a landing into the wind. The first few times I practiced downwind landings, it amazed me how fast I seemed to be going as I prepared to flare, even though the airspeed indicator displayed normal values. For this reason you will need to work hard at correcting for the additional ground speed the plane will be carrying through the flare and touchdown. If you do not, the plane will carry much farther down the runway before it finally touches down.

Depending on the length of the runway you are landing on, it might be necessary to touch down very close to the approach end of the runway, once again requiring the use of many of the landing techniques discussed throughout this book. Any excess airspeed carried on final will result in additional runway being necessary to stop. If the wind is a crosswind, remember that you might need to change the use of controls to prevent the wing from being raised once you are on the ground. Soon after touching down, depending on the strength of the crosswind, it might be necessary to revert to the tailwind taxiing techniques that were discussed in chapter 8.

Use of downwind landings will also increase the wear on your brakes and tires, especially when making short-field landings. If you are making a soft-field landing, the additional ground speed might increase the tendency of the plane to nose over for both conventional- and tricycle-gear aircraft. You should also use your airplane's performance charts to determine the runway length that will be necessary to get down and stopped for the conditions you are flying in.

Downwind landings can be made safely if you know what to expect as you make the approach, and they can be a useful option if conditions do not permit the standard approach to landing. Before you attempt downwind landings, it is a good idea to find a flight instructor to work with you.

SUMMARY

We have reviewed a variety of emergency landing topics in this chapter, but we have not even scratched the surface of every situation that could take place. However, we have attempted to cover some of the basics that can be useful in almost any emergency landing. Knowing your plane's emergency procedures and practicing them will reduce the level of tension in a true emergency. As stated many times in this chapter, you should keep your head when an emergency arises and maintain control of the plane. In talking with FAA personnel, the reason for many crashes in emergency situations is that the pilots are so focused on the emergency itself, they forget to keep flying the plane.

By remaining calm and thinking your way through the emergency, you won't forget steps in the emergency procedures. Sometimes an emergency landing can be avoided entirely if you follow the proper procedures. For instance, many forced landings are the result of fuel starvation as one tank runs empty and the pilot fails to switch to a tank that has fuel remaining. By going through the emergency procedures, you will take this step, and the engine will normally restart and allow you to continue your flight. Carburetor ice is another example. By applying carburetor heat, it is often possible to get the engine running again if it has stopped due to carburetor ice. Be aware that when you begin to experience engine problems, you should apply carb heat as soon as possible. The heat that warms the air as carb heat is applied comes from the engine itself. The engine will begin to cool rapidly as the engine loses power. If you wait too long to apply carburetor heat, there might not be sufficient warmth to heat the air and melt any ice. Both of the just-mentioned steps are normally in the emergency-landing procedures for a plane. By following them you can avoid a true emergency and still have an interesting story to tell your flying buddies, as opposed to ending up in an FAA accident report.

12
Conclusion

I HAVE BEEN FASCINATED WITH FLYING FOR A VERY LARGE PORTION OF MY life. As a flight instructor, I have attempted to share this fascination and to teach my students how to fly as safely as possible. This desire to have pilots fly as safely as they can, within their flying capabilities, extends to takeoffs and landings. Much of the material presented in this book is the same material you were taught as a student pilot, and you might have covered the lessons again during your biennial flight reviews.

In this conclusion we will cover a number of topics that relate to flight safety and pilot proficiency. Even if you have thousands of hours logged, you should always look for ways to continue to improve your understanding and awareness of these two topics. There are a number of sources available to you to aid in reaching that goal. Among them are ongoing flight instruction, practice on your own, FAA safety seminars, and the Wings Program. We will discuss each of these topics in this section, in addition to covering the poor judgment chain and mental processes and attitudes you should be aware of as you fly.

WHERE TO GO FROM HERE

Having read this book, you should now have a good understanding of what the procedures are for each takeoff and landing scenario, and now you are wondering how to put the information to practical use. There are several options open to you for building on your past experience and improving your level of proficiency and safety. In this section we are going to discuss flight instruction from the standpoint of training reviews and aerobatics training, in addition to several of the topics previously mentioned. Let's begin with flight instruction.

Flight instruction

I have always felt that it is a good idea for pilots to take an occasional refresher course in the various maneuvers they might need to perform. As a rule, most pilots fly to the same airports and establish a set of patterns in how they fly. They know that the runway at airport A has a dip in the middle, so they try to make sure the plane is firmly planted before they reach it. Airport B has a bubble of air over the trees if the wind is from the right direction, and pilots are prepared for the rolling motion it might impart to the plane as they take off or land. When flying between these frequent airport haunts, pilots fly straight and level at cruise speed, rarely, if ever, doing a stall, a steep-banked turn, or a flight at minimum controllable airspeed.

For this reason it is a good idea for these very experienced, capable aviators to get a flight instructor to fly along once in a while to run through some of these maneuvers. Not only does this increase pilots' proficiency and safety, but it also lets them retain their "feel" for the plane in these less frequently used flight regimes. While the maneuvers themselves are not inherently hard or dangerous, it is a good idea to have a flight instructor along every now and again to observe and critique your performance. Just by listening to what other pilots and flight instructors have to say, it amazes me how many times I can learn something that helps me to better understand how to execute a particular maneuver or increase the safety of a given situation.

Using a flight instructor to help you review areas in which you are rusty should not be considered a negative reflection of your flying abilities, but a positive sign of your desire to improve the proficiency and accuracy with which you fly. The biennial flight review is intended to test a pilot's proficiency level in flight regimes they do not get much practice in, but doing a short-field landing every two years can hardly be considered enough practice to do more than just brush up on some of the highlights. A qualified flight instructor can help you understand how to correctly fly each maneuver in the prescribed manner. Find a flight instructor you are comfortable with, and head out for a little training review once or twice a year. It will help keep some of the take-off, landing, and flight maneuvers fresh in your mind.

Unusual attitudes/aerobatics. A fair amount of this book has been dedicated to reviewing the emergency procedures related to takeoffs and landings. Another area that you should consider recurring training in is unusual attitudes. This should include both VFR and simulated IFR conditions. In these training sessions, the instructor will put the plane into unusual attitudes, normally while the student is looking at the floor of the plane, then turn it over to the student to recover from the unusual attitude. The reason behind this type of training is to maintain proficiency in reacting to these types of unusual situations.

At this point you might be asking yourself, "How can a VFR pilot get into an unusual attitude, since they are normally watching outside the airplane?" Every so often, pilots become distracted inside the cockpit. This can come in the form of dropping a chart or some other article onto the floor of the plane, talking to a passenger, or many other reasons. For a short period of time, they forget about flying, and when they look up, the plane is in an unusual attitude. For pilots who have never been in an attitude of an extremely high, nose-up, 75° bank or a spin, this can come as a startling situation. Under normal circumstances, unusual attitudes are easy to recover from, but it is best to get instruction from a qualified flight instructor once or twice a year to maintain your proficiency.

Related to unusual attitudes is aerobatics training. I have flown aerobatics for years and believe that pilots who have received proper aerobatics training are less likely to panic if they find themselves in an unusual attitude. Wake turbulence behind large aircraft has been known to roll smaller aircraft into the inverted position. To untrained pilots, attempting a half loop to upright seems like the best way to recover. Most of the time they do not have sufficient altitude and do not make it. A roll is the quickest way back to upright flight in most situations, but the pilot, who has attempted to keep the plane from rolling in the first place, finds it difficult to change gears mentally at that point.

Every so often, pilots who have never been in a spin end up in one, not even realizing it. Today, training regulations do not require spin training for most pilot certificates, and even those that do are inadequate. Spin training from a qualified flight instructor, in the proper aircraft, can help pilots understand how to get out of spins and how they actually get into them. For many pilots, spins represent a wall they are not interested in climbing, and this should not be the case. An hour or two of spin training from the right flight instructor, broken up in intervals, will make you a safer pilot.

In addition to spin training, taking a few hours of aerobatics instruction can also be a confidence builder. Knowing how to do rolls, loops, spins, and a few other basic aerobatic maneuvers will help you in several ways. First, you will have experience and will be more comfortable in extremely unusual attitudes. Seeing the horizon roll around the nose of the plane the first time you do a roll will amaze you. Knowing where to look for reference points during a loop will teach you how to know what the plane's attitude is during the execution of the maneuver. Second, you begin to understand that planes can fly under these conditions. I have talked to aerobatics students that were initially unsure that the plane would continue to fly during aerobatics. As they practice flying maneuvers, they begin to understand that aerobatics planes fly equally well upside down as right-side up. Third, you begin to learn how to fly a plane by feel. During aerobatics you are looking outside the cockpit. You are constantly craning to see the horizon during maneuvers, looking left, right, over your head, and behind you to monitor your attitude. After a while you know the airspeed by the sounds the plane is making and the control pressures. You don't think about how you should move your hands and feet. You blend yourself with the plane to make it do what you want.

All of this makes you a safer, more competent pilot. Having been inverted, you are less intimated by it. The same holds true with spins or rolls. But you need to learn from qualified instructors in the right types of aircraft. Proper aerobatics training begins with an airplane that is certified in the aerobatics category. To fly aerobatics, you must wear a parachute that has been properly packed within the specified time period for that chute type. There are also several other regulations related to aerobatics, altitude, and visibility.

You might find it hard to find flight schools in your area that teach spins, let alone aerobatics, but it is worth the time to find a good school and take the training. The International Aerobatics Club, in Oshkosh, Wisconsin, maintains a list of aerobatics flight schools. By contacting this organization, you should be able to find an aerobatics flight school near you. As you talk to an aerobatics flight school, you should get the feel that safety is their primary goal. If you can find a school with a flight instructor who competes in aerobatics, so much the better. Participating in aerobatics competitions, while allowing friendly competition between pilots, also shows that the pilot is interested in safety and precision in flying maneuvers. This safety and precision will also manifest itself in the aerobatics training you receive, ultimately making you a safer pilot.

Practice

Once you are proficient in the various takeoffs and landings, in addition to other flight maneuvers, you should practice them on a regular basis. This will help you keep your proficiency and improve your flying abilities. Too often, pilots get into the habit of flying the same takeoffs and landings from the same airports, and their abilities atrophy in other areas. Maintain your abilities by practicing as frequently as you can. When I fly, I like to challenge myself to do each thing as well as I possibly can. Before a flight, I will set a goal I want to accomplish and work at that. This helps me work on areas that I need practice in and increases my enjoyment of flying. By setting goals for yourself, you can strengthen any areas you might need practice in. The most important thing is to practice and fly safely. Remember, if you are at all unsure, take a qualified flight instructor along to help you get up to speed on a particular maneuver.

Safety seminars

The FAA, Department of Transportation, and other federal and state agencies frequently offer flight safety seminars on a regular basis. The topics discussed cover a wide range of aviation-related areas and can give you exposure to information that you might not normally have found. I recently attended an aerobatics safety seminar, and the speakers were extremely good. Topics ranged from inverted oil systems, to how to fly certain maneuvers, to emergency spin recovery. Other seminars can cover topics such as regulation changes, cross-country flying, emergency situations, and other areas of interest. By attending such seminars, you are exposed to valuable information, and you can learn about what these agencies provide for you. There is a great deal of information they can send you that is available for the asking, and they are normally very willing to help you with questions or concerns you might have. Taking advantage of the safety seminars can help you be a safer pilot.

Wings Program

The FAA has a program known as the Wings Program. This is a series of seminars and flight training that is intended to help pilots increase their proficiency, knowledge, and flight safety. In some cases, by complying with the rules of the Wings Program, you can avoid the need for a biennial flight review, using the program as an ongoing method of meeting those requirements. You should contact your local FAA flight standards district office (FSDO) to find out more about the program that is being run in your area.

In many cases, some of the seminars mentioned in the previous section are also part of the Wings Program, and in others there are more specific requirements that you must meet to participate. But if you are interested in continuing your aviation education, this can be a worthwhile program to get into.

FLIGHT SAFETY

Throughout the book we have discussed how important flight safety is. In this section we are going to review three topics related to the mental processes associated with safety. This information comes from FAA publications and includes the poor judgment chain, three mental processes of safe flight, and five hazardous attitudes. Reports show that many aviation accidents are avoidable. They are often precipitated by poor judgment calls on the part of the pilot. Schedule pressures, fatigue, not knowing limitations, and other factors can cause pilots to take risks that are unnecessary and begin a series of events that end with an accident. By interrupting that chain of poor mental steps, you can avoid many accidents. We will discuss these in the following sections.

Poor judgment chain

Publication FAA-P-8740-53, "Introduction to Pilot Judgment," discusses the poor judgment (PJ) behavior chain. As previously stated, most accidents are caused by a series of events. Here is an example used in this FAA document. A noninstrument-rated pilot is under a schedule restraint and is running late. He has limited flight experience in adverse weather, yet decides to fly through an area of possible thunderstorms just before dark. He does not trust his instruments, and the darkness, turbulence, and heavy clouds cause him to become disoriented as a result.

You can see that several factors went into this situation. Time pressures, lack of experience in flying in poor weather, and instrument flying all set this pilot up for a bad situation. If the pilot had used good judgment at several points in the flight, he could have avoided this. First, flying around the weather and being less concerned about being late would have avoided the situation entirely. Second, he could have avoided flying into the turbulence and clouds once he did encounter the weather. Finally, once in them he could have trusted his instruments and not become disoriented ("Introduction to Pilot Judgment," p. 1).

There are two major principles to the poor judgment chain. These are the following: One poor judgment increases the probability that another will follow. Judgments are based on information pilots have about themselves, the aircraft, and the environment, and pilots are less likely to make a poor judgment if this information is accurate. Thus, one poor judgment increases the availability of false information, which might then negatively influence judgments that follow ("Introduction to Pilot Judgment," p. 3).

As the PJ chain grows, the alternative for safe flight decreases. If a pilot selects only one alternative among several, the option to select the remaining alternatives might be lost. For example, if a pilot makes a poor judgment and flies into hazardous weather, the alternative to circumnavigate the weather is automatically lost ("Introduction to Pilot Judgment," p. 3).

As you can see, pilot judgment plays a crucial role in flight safety. By interrupting the poor judgment chain early, the pilot is left with more options for a safe flight. By delaying the use of good judgment, the pilot might reach a point at which there are no good alternatives. Make it a habit of making the best decisions you can and avoid the negative side effects of poor decisions ("Introduction to Pilot Judgment," p. 3).

Three mental processes of safe flight

The "Introduction to Pilot Judgment" document also covers three mental processes related to safe flight. These are automatic reaction, problem resolving, and repeated reviewing. In this section we will briefly discuss each one of these areas. Let's begin with automatic reaction.

Good pilots can perform many of the activities associated with flying a plane automatically. Attitude, heading, and altitude control can all become automatic functions that pilots perform once they have enough experience in flying a particular plane. Reaction to some emergency situations can also become automatic, with the pilots' training taking over and getting them through it. As we discussed in the aerobatics section, after a period of time, pilots no longer think about how they need to move the controls during a certain maneuver. Instead, they automatically take the proper actions to make the plane fly the way they want it to. This is known as automatic reaction. You are flying the plane without conscious effort, and good pilots need to achieve this capability in certain aspects of flying.

Second on the list is problem solving. This is also necessary to become a safe pilot, and it is a three-step process that includes:

1. Uncover, define, and analyze the problem.
2. Consider the methods and outcomes of possible solutions.
3. Apply the selected solution to the best of your ability.

By taking these steps, you will improve your ability to understand and resolve problems that arise as you fly. Correctly determining the actual cause of a problem, rather than misunderstanding it, can aid in making better decisions in resolving it. As you can see, in some cases the poor judgment chain can be broken through the use of good problem-solving techniques.

The third and last mental process is repeated reviewing. This is the process of "continuously trying to find or anticipate situations which might require problem resolving or automatic reaction." Part of this includes using feedback related to poor judgment chains. As you fly, you need to constantly be aware of the factors that affect your flight, including yourself, the plane, and the weather. By remaining aware of the entire situation, you will be more highly informed and better able to analyze the actual conditions you are flying under ("Introduction to Pilot Judgment," p. 5).

Five hazardous attitudes

The last topic we will review from the "Introduction To Pilot Judgment" are the five attitudes that can be hazardous to safe flight. These include:

- Antiauthority: "Don't tell me!"
- Impulsivity: "Do something—quickly."
- Invulnerability: "It won't happen to me."
- Macho: "I can do it."
- Resignation: "What's the use?"

Each of these attitudes can reduce pilots' abilities to make sound judgments as they fly. Ego, part of any pilot's makeup, can get in the way of taking the safest course of action in a given situation. Exhibiting these five attitudes increases the chances that the poor judgment chain will not be broken. Monitor yourself. If you catch yourself exhibiting one or more of the attitudes described, work to control it and not let it control you.

We have briefly reviewed some very important points that were part of the "Introduction to Pilot Judgment" document (p. 10). I highly recommend that you contact your local FAA office and ask them to send you a copy of it. There are a number of exercises in it to help you better understand each of the areas we covered in this section. Maintaining the right attitude and using good judgment are the best things pilots can do to increase the safety of their flights.

SUMMARY

This book has covered a number of topics related to takeoffs, landings, and related topics. The message of flight safety should be clear. Taking unnecessary risks will catch up with pilots eventually, and you have the power to avoid many dangerous situations by merely exercising good judgment. Takeoffs and landings are some of the most important flight maneuvers you will perform every time you fly. You should approach them in a professional manner, attempting to improve every one you do, with perfection as the goal. While no one can ever achieve this level of proficiency, setting it as the goal will help you to constantly improve. This desire for perfection will raise your level of expectations in all areas of flight-related activities and help to improve your overall piloting abilities.

Always remember that you are flying the plane. Within the plane's and your capabilities, make it fly the way you want it to. Don't let your plane fly you. Critique yourself on every flight and look for ways to improve. Flying should be one of the most enjoyable activities you perform. By remaining proficient, practicing the various take-offs and landings, and using flight instructors when necessary, you increase your probability of ending every flight safely.

Bibliography

Airman's Information Manual. (Revised 1993) Federal Aviation Administration, Washington D.C.

Airport/Facility Directory. (6 January 94) U.S. Department of Commerce, Washington D.C.

Flight Training Handbook. (Revised 1980) Federal Aviation Administration, Washington D.C.

"Impossible Turn." Accident Prevention Program. Federal Aviation Administration, Washington D.C.

"Introduction To Pilot Judgement." Accident Prevention Program. Federal Aviation Administration, Washington D.C.

"On Landings: Part I." Accident Prevention Program. Federal Aviation Administration, Washington D.C.

"On Landings: Part II." Accident Prevention Program. Federal Aviation Administration, Washington D.C.

"On Landings: Part III." Accident Prevention Program. Federal Aviation Administration, Washington D.C.

Pilot's Handbook of Aeronautical Knowledge. (Revised 1980) Federal Aviation Administration, Washington D.C.

Index

A

abort procedures, 183-193
  landings, 187-192
  takeoffs, 184-187
aerobatics, 220-222
aerobraking, 68
air density, 10-11
aircraft (*see also* conventional-gear aircraft; tri-cycle-gear aircraft)
  flying in winter, 167-181
  frost/snow on, 173-175
  ice on, 174-175
  manufacturer's compromise, 90-91
  preflight, 3-7
  separation, 42-43
    radio sequencing, 45
    visual spacing, 45
airfoil, 69
Airman's Information Manual (AIM), water ditching, 211-212
airports, ground markings, 28-30
airspeed
  control for landing, 54-55
  correction table, 19
  crosswind, 151-153
airspeed indicator, 18-20
altitude
  density, 104-105
  pressure, 104
angle of attack, 15
  critical, 16
approaches (*see also* landings)
  flat, 74-75
  initial radio call, 23-25

B

ballooning, 14, 73-74

C

cabin heaters, 172
camber, 68
charts
  landing, 109-113
  performance, 104-105
  takeoff, 105-109
checklists
  complex single-landing, 53
  GUMP, 53
  landing, 51-53
  trainer landing, 52
cockpit, preflight inspection, 4
control flutter, 17
conventional-gear aircraft
  flare/rollout, 98-99
    for crosswind landings, 161-165
    for short-field landings, 121-125
    for soft-field landings, 141-143
  ground roll/rotation, 91-94
    for crosswind takeoffs, 155-158
    for soft-field takeoffs, 133-136
  touchdowns, 62-67
chord line, 15
crab methods, 164-165
crosswind landings, 152, 159-166
crosswind takeoffs, 153-159

D

defrosters, 172
density altitude, 104-105
departures, 25-26
detonation, 199
drag, 70
  form, 70
  induced, 70
  parasite, 70
  skin-friction, 70

E

emergencies
  abort procedures, 183-193
  downwind landings, 215-217
  engine loss on approach to landing, 204-207
  engine loss on takeoff, 200-204
  flat-tire landings, 214-215
  forced landings, 204
  instrument scanning, 196-197
  landing on water, 210-213
  landing procedures, 195-217
  retractable-gear problems, 213-214
  rugged-terrain landings, 207-209
  warning signs, 197-200
engine offset, 90
engines
  loss on approach to landing, 204-207
  loss on takeoff, 200-204
  preheating during winter, 168-171
  starting during winter, 171-172
  warning signs, 198-199

F

Federal Aviation Regulation (FAR), Part 61.57, 1-2
flaps
  designs, 85
  retracting, 86
  setting for landings, 96
    crosswind, 160-161
    short-field, 119-120
    soft-field, 140
  setting for takeoffs, 84-86
    crosswind takeoffs, 154-155
    short-field takeoffs, 113-118
    soft-field takeoffs, 131-132
flares/flaring, 62-68
  conventional-gear aircraft, 98-99
    for crosswind landings, 161-165
    for short-field landings, 121-125
    for soft-field landings, 141-143
  height judgment during, 61-62
  night landings, 77, 79
  tricycle-gear aircraft, 99-100
    for crosswind landings, 165-166
    for short-field landings, 125-127
    for soft-field landings, 143-144
  vision during, 60-61
flight
  night, 76-81
  over mountain terrain, 197
  winter, 167-181

flight instruction, 220-222
form drag, 70
frost, 173-175
fuel, 199

G

glideslope
  obstacle clearance, 122
  VASI, 34
ground effect, 68-71, 130-131
ground markings, 28-30
ground roll
  conventional-gear aircraft, 91-94
    crosswind takeoffs, 155-158
    soft-field takeoffs, 133-136
  tricycle-gear aircraft, 94-95
    crosswind takeoffs, 158-159
    soft-field takeoffs, 136-139
    short-field takeoffs, 118-119
groundloops, 75-76
gyroscopic precession, 89-90

H

high-intensity runway lights (HIRL), 80
humidity, 10-11
hydroplaning, 81

I

ice, 174-175
  aircraft skidding on, 179
induced drag, 70
inspection, preflight aircraft, 3-7

L

landings, 51-82, 95-100
  abort procedures, 187-192
  airspeed/power control, 54-55
  ballooned, 73-74
  bounced, 71-73
  checklists, 51-53
    complex single-, 53
    trainer, 52
  conventional-gear aircraft flare/rollout, 98-99
    for crosswind, 161-165
    for short-field, 121-125
    for soft-field, 141-143
  crosswind, 152, 159-166
  downwind, 215-217
  during winter weather, 178-179
  emergency procedures, 195-217
  engine loss on approach, 204-207

flap settings, 96
    for crosswind, 160-161
    for short-field, 119-120
    for soft-field, 140
flat approaches, 74-75
flat-tire, 214-215
forced, 204
full-stall, 65-67
ground effect, 68-71
groundloops, 75-76
height judgment during flare, 61-62
high final approach, 58
low final approach, 59
night, 76-81
    approach planning, 76-77
    flaring, 77, 79
    height judgment, 77, 79
    obstacle clearance, 80-81
normal final approach, 59
on rough ground, 208-209
on trees, 207-208
on water, 210-213
picking touchdown point, 55-60
preparing charts for, 109-113
retractable-gear problems, 213-214
rollout, 67
rudder settings, 97-98
rugged-terrain, 207-209
short-field, 119-127
skidding on ice, 179
soft-field, 139-144
touchdown, 62-68
tricycle-gear aircraft flare/rollout, 99-100
    for crosswind, 165-166
    for short-field, 125-127
    for soft-field, 143-144
trim settings, 96-97
    for crosswind, 161
    for short-field, 120-121
    for soft-field, 140
undershoot, 74
unusual situations, 71-74
vision during flare, 60-61
wet runways, 81
wheel, 64-65
landing tee, 29-30
lighting
    runway, 79-81
    VASI, 34-35
low-intensity runway lights (LIRL), 80

M

medium-intensity runway lights (MIRL), 80

O

obstacles
    clearing at night, 80-81
    runway, 80-81

P

P-factor, 87-88
parasite drag, 70
performance charts, 104-105
pilot proficiency, 1-3
    definition/description, 2
    flight frequency and, 2-3
pilots
    poor judgment chain, 223-224
    routine instrument scans, 196-197
preflight inspection
    aircraft, 3-7
    cockpit, 4
    cowling left side, 5
    front of aircraft, 4-5
    left wing, 5-6
    right wing, 7
    tail surfaces, 6
pressure altitude, 104

R

relative wind, 15
rollout
    conventional-gear aircraft, 98-99
        for crosswind, 161-165
        for short-field, 121-125
        for soft-field, 141-143
    tricycle-gear aircraft, 99-100
        for crosswind, 165-166
        for short-field, 125-127
        for soft-field, 143-144
rudder
    crosswind control positioning, 150-151
    settings for landing, 97-98
    settings for takeoff, 87
runways
    designation marking, 30
    fixed-distance marker, 30
    lighting, 79-81
    markings, 30-34
    night-landing view, 78
    precision-instrument, 33
    segmented circle/intersecting, 32
    segmented circle/pattern, 31
    simple markings, 32
    taxiway hold lines, 33

runways *continued*
  visual approach slope indicator, 34
  wet, 81

S

safety, 1-11
  abort procedures, 183-193
  aircraft preflight, 3-7
  attitudes for hazardous flight, 225
  attitudes for safe flight, 224
  emergency-landing procedures, 195-217
  flight, 223-225
  pilot proficiency and, 1-3
  seminars on, 222
  weather, 7-11
segmented circle, 30
short-field landings, 119-127
short-field takeoffs, 113-119
skin-friction drag, 70
skydivers, 198
slipstream, 89
snow, 173-175
soft-field landings, 139-144
soft-field takeoffs, 131-139
speed (*see* airspeed; V-speeds)
stalls, 130-131 (*see also* engines)
  recovery, 14-15
  V-speeds and, 13-14

T

tail, inspecting, 6
takeoffs, 84-95
  abort procedures, 184-187
  climb rudder, 87
  conventional-gear aircraft ground roll/rota-
    tion, 91-94
    for crosswind, 155-158
    for soft-field, 133-136
  crosswind, 153-159
  crosswind control positioning, 150-151
  during winter weather, 178
  engine loss during, 200-204
  flap settings, 84-86
    for crosswind, 154-155
    for short-field, 113-118
    for soft-field, 131-132
  gyroscopic precession, 89-90
  P-factor, 87-88
  preparing charts for, 105-109
  short-field, 113-119
  slipstream, 89
  soft-field, 131-139

torque, 88-89
traffic patterns, 46-47
tricycle-gear aircraft ground roll/rotation, 94-
  95
  for crosswind, 158-159
  for short-field, 118-119
  for soft-field, 136-139
trim setting, 86-87
  for crosswind, 155
  for soft-field, 132-133
taxiing
  crosswind, 148-149
  crosswind control positioning, 150-151
  winter runway conditions, 176-177
taxiways, hold lines, 33
temperature, 10-11, 104
tires, 175-176
  flat, 214-215
torque, 88-89
touchdowns, 55-60, 62-68
  45-degree point, 56-60
  conventional-gear aircraft, 62-67
  skidding on ice, 179
  tricycle-gear aircraft, 67-68
  undershoot, 74
traffic patterns, 23-49
  aircraft separation, 42-45
  airport/facility directory listing, 26-27
  base entry, 39
  busy, 46-47
  crosswind entry, 37
  crosswind exit, 40-41
  departures, 25-26
  downwind entry, 38-39
  entry, 34-40, 46
  exit, 40-42
  final entry, 39-40
  ground markings, 28-30
  initial radio call approaches, 23-25
  left, 26-28
  pattern direction indicators, 27
  pattern legs, 26
  pattern wind correction, 47-48
  right, 26-28
  right-turn exit, 41-42
  runway markings, 30-34
  straight-out exit, 42
  takeoffs, 46-47
tricycle-gear aircraft
  flare/rollout, 99-100
    for crosswind landings, 165-166
    for short-field landings, 125-127
    for soft-field landings, 143-144

ground roll/rotation, 94-95
  for crosswind takeoffs, 158-159
  for short-field takeoffs, 118-119
  for soft-field takeoffs, 136-139
  touchdowns, 67-68
trim
  setting for landings,
    crosswind, 161
    short-field, 120-121
    soft-field, 140
  setting for takeoffs, 86-87
    crosswind, 155
    soft-field, 132-133
turbulence, wake, 43-45 (*see also* wind shear)
turning, 88

V

V-speeds, 13-21
  best angle of climb, 17
  best rate of climb, 17-18
  in the pattern, 20
  maneuvering speed, 16-17
  maximum flap-extended speed, 17
  never-exceed speed, 17
  stall recovery, 14-15
  stall speed clean configuration, 14
  stall speed landing configuration, 13-14
vertical stabilizer, 90
visual approach slope indicator (VASI), 34
vortex, wingtip, 43-45

W

wake turbulence, 43-45
washout, 90-91
water, landing on, 210-213
weather, 7-11
  frost/snow on aircraft, 173-175
  ice on aircraft, 174-175
  temperature/humidity, 10-11
  wet runways, 81
  wind shear, 8-9
wheelpants, 175-176
wind
  pattern correction and, 47-48
  relative, 15
wind shear, 8-9
wind sock, 28
wind tetrahedron, 29
wings, inspecting, 5-7
Wings Program, 223
winter flying, 167-181
  cabin heater/defroster, 172
  frost/snow on aircraft, 173-175
  ice on aircraft, 174-175
  landings, 178-179
  preflight routine, 168
  preheating the engine, 168-171
  starting the engine, 171-172
  takeoffs, 178
  taxiing, 176-177
  tires/wheelpants, 175-176

About the author

Michael Charles Love flies on weekends for a local skydiving organization in a Cessna 182. Due to the nature of the flying, it is not unusual for him to execute 10 or more take-offs and landings in a single day—often in strong crosswind conditions from a small, grassy runway. It's possible that this is the origin of *Better Takeoffs and Landings*.

It's hard to judge, though. Love has been taking off and landing since the age of 15. He holds commercial, instrument, and CFI ratings, and airframe and power-plant mechanic ratings. He is a member of the Experimental Aircraft Association, the International Aerobatic Club, and the Madison (WI) Astronomical Society. Love gives aerobatic flight instruction and participates in aerobatic competitions using a Pitts Special S-2B, a conventional-gear plane that is particularly demanding to land. He holds several college degrees, including an M.S. in Computer Science, a B.S. in Aviation Management, an A.A.S. in Aviation Maintenance, and an A.A.S. in Data Processing. When he is on the ground, Love works in the computer industry.

Other Bestsellers of Related Interest

ABCs of Safe Flying, 3rd Edition
—David Frazier
This clearly written pilot's guide describes the keys to competency as a pilot, the habits of safety, and the attitude of the professional. Also covers ground operations, airspace designations, Federal Aviation Regulations, naviational equipment, cross-country flights, pilot/controller cooperation, and aviation career opportunities.
ISBN 0-8306-2089-3 $15.95 Paper
ISBN 0-8306-2091-5 $22.95 Hard

Art Of Instrument Flying, 2nd Ed.
—J. R. Williams
A comprehensive guide to all elements of IFR flight, with review questions and answers at the end of many chapters to reinforce the subject matter.
ISBN 0-8306-3654-4 $21.95 Paper
ISBN 0-8306-7654-6 $31.95 Hard

Pilot's Radio Communications Handbook (The), Revised 4th Edition
—Paul E. Illman
Everything VFR pilots need to know to communicate from the cockpit effectively and use even the busiest airports with confidence. Now updated and expanded to cover new U. S. airspace designations.
ISBN 0-07-031756-9 $17.95 Paper
ISBN 0-07-031757-7 $29.95 Hard

Stalls & Spins
—Paul Craig
A practical guide to help private pilots and flight students meet new FAA requirements, this book demystifies stalls and spins. Readers gain a better knowledge and understanding of the aerodynamic principles involved, the psychological effects of stalling and spinning, the actions necessary to avoid disaster, and the spin characteristics of specific aircraft.
ISBN 0-8306-4020-7 $18.95 Paper
ISBN 0-8306-4019-3 $26.95 Hard

How to Order

Call 1-800-822-8158
24 hours a day,
7 days a week
in U.S. and Canada

Mail this coupon to:
McGraw-Hill, Inc.
P.O. Box 182067,
Columbus, OH 43218-2607

Fax your order to:
614-759-3644

EMAIL
70007.1531@COMPUSERVE.COM
COMPUSERVE: GO MH

Shipping and Handling Charges

Order Amount	Within U.S.	Outside U.S.
Less than $15	$3.50	$5.50
$15.00 - $24.99	$4.00	$6.00
$25.00 - $49.99	$5.00	$7.00
$50.00 - $74.49	$6.00	$8.00
$75.00 - and up	$7.00	$9.00

EASY ORDER FORM— SATISFACTION GUARANTEED

Ship to:

Name _____

Address _____

City/State/Zip _____

Daytime Telephone No. _____

Thank you for your order!

ITEM NO.	QUANTITY	AMT.

Method of Payment:

☐ Check or money order enclosed (payable to McGraw-Hill)

☐ VISA   ☐ DISCOVER

☐ AMERICAN EXPRESS Card   ☐ MasterCard

Shipping & Handling charge from chart below	
Subtotal	
Please add applicable state & local sales tax	
TOTAL	

Account No. ☐☐☐☐☐☐☐☐☐☐☐☐☐☐☐☐

Signature _____ Exp. Date _____
Order invalid without signature

In a hurry? Call 1-800-822-8158 anytime, day or night, or visit your local bookstore.

Code = BC15ZZA